J.W. Niemantsverdriet, J.P.K. Felderhof

Scientific Leadership

## Also of interest

*Productivity and Organizational Management*
Machado, Davim (Eds.), 2017
ISBN 978-3-11-035545-1, e-ISBN: 978-3-11-035579-6

*Innovation Technology*
*A Dictionary*
Schramm, 2017
ISBN 978-3-11-043824-6, e-ISBN 978-3-11-042917-6

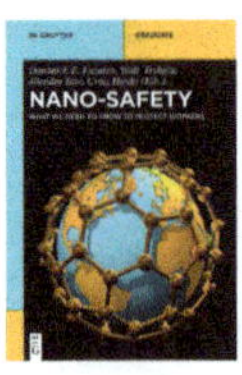


*Nano-Safety*
*What We Need to Know to Protect Workers*
Fazarro, Trybula, Tate, Hanks (Eds.), 2017
ISBN 978-3-11-037375-2, e-ISBN 978-3-11-037376-9

*The Science of Innovation*
*A Comprehensive Approach for Innovation Management*
Löhr, 2016
ISBN 978-3-11-034379-3, e-ISBN 978-3-11-034380-9

*Engineering Risk Management*
Meyer, Reniers; 2016
ISBN 978-3-11-041803-3, e-ISBN 978-3-11-041804-0

J.W. Niemantsverdriet, J.P.K. Felderhof

# Scientific Leadership

DE GRUYTER

**Authors**
Prof. J.W. Niemantsverdriet
SynCat Scientific Leadership BV
and Syngaschem BV
Nuenen, The Netherlands
jwn@syngaschem.com

J.P.K. Felderhof
SynCat Scientific Leadership BV
and Impetus BV
Eindhoven, The Netherlands
jan-karel@syngaschem.com

ISBN 978-3-11-046888-5
e-ISBN (E-BOOK) 978-3-11-046889-2
e-ISBN (EPUB) 978-3-11-046921-9

**Library of Congress Cataloging-in-Publication Data**
A CIP catalog record for this book has been applied for at the Library of Congress.

**Bibliographic information published by the Deutsche Nationalbibliothek**
The Deutsche Nationalbibliothek lists this publication in the Deutsche Nationalbibliografie; detailed bibliographic data are available on the Internet at http://dnb.dnb.de.


Cover image: Ginosphotos/iStock/Thinkstock
Typesetting: Compuscript Ltd. Shannon, Ireland
Printing and binding:
∞ Printed on acid-free paper
Printed in Germany

www.degruyter.com

# Foreword

This book is the result of a mutually inspiring collaboration of over 25 years. The authors met for the first time in the early nineties: Jan Karel was already established as a leadership development coach with programs in the Eindhoven area with views on how to transform authenticity in performance and impact; Hans was an associate professor at the Eindhoven University of Technology (TU/e), who had just built up an academic research group that was beginning to see some successes. They never lost contact ever since, and when Hans assumed an executive role at the university in 2001, he invited Jan Karel as his personal coach, to be better prepared for the new "Deanly Duties."

The cooperation intensified in 2013, when Yong-Wang Li, founding manager of Synfuels China, invited Hans to help build up a new, internationally oriented laboratory for fundamental research, SynCat@Beijing, with a local branch in the Netherlands, Syngaschem BV, and be the director of both. Jan Karel joined as the Director Strategy, Organization and Human Resources, of Syngaschem BV. Together, they developed a course program on scientific leadership for the newly appointed scientific staff of SynCat, in the framework of the SynCat Ac@demy.

The authors organized five sessions, two at the Boordhuys in Nuenen, the Netherlands, and three at Synfuels China's headquarters in Beijing-Huairou. Several guest speakers made excellent contributions: Yong-Wang Li, Jens Rostrup-Nielsen, Michael Bowker, and Gilbert Froment.

This book is an extended version of the program that the authors developed for the SynCat Ac@demy (www.scientificleaders.com). It finds its roots in the expertise of Jan Karel in (self)-leadership since the 1990s in his company Impetus BV and the programs on presenting science that Hans developed at TU/e in the period 1990–2010.

https://doi.org/10.1515/9783110468892-202

# Acknowledgments

The authors are grateful to several colleagues, mentors, and friends with whom they developed ideas and course programs, which have become essential ingredients in the philosophy behind this book.

Jan Karel Felderhof wishes to thank the following:

- Dr. Martin Bakker, director at Océ Research, Venlo; together, we designed a program on how to integrate technology development, product development, and product engineering, from the viewpoint of both research and relevance in the market.
- Ir. Piet Spohr, director of the TNO Physical and Electronical Laboratory; we developed ideas on how to focus a research institute of 400 scientists and engineers on projects and markets, on the basis of own science and technology.
- Ir. Poul Bakker, chairman of the Company Coaching at Philips; together we set up talent programs, in particular for the Philips Research Institute "NatLab."
- Professor Anton Hemerijck, director of the Dutch Scientific Council for Government Policies (WRR), dean at the Free University of Amsterdam, and professor at the London School of Economics, for collaboration on formulating theories for building stronger organizations on the basis of trust, relevance, and self-confidence.
- Prof. Dr. Piet Cijsouw, director of the TU/e Post Graduate School, where, together, we designed professional development programs for alumni and others.

Hans Niemantsverdriet is grateful to the following:

- Diek Koningsberger, who stood at the basis of the jointly developed course "Oral and Poster Presentations" for PhD and MSc students at TU/e around 1990. The course was later published as a brochure and distributed through the channels of the European Federation of Catalysis Societies (EFCATS), both in print and online. It is incorporated in Chapter 3.
- Roel Prins (ETH Zurich) developed excellent lectures on how to write scientific papers. Around 2010, Roel and Hans joined forces and made a combined video course, with the highly appreciated support of Elsevier's Lily Khidr and TU/e's Maurice Megens. It is still on the internet (www.catalysiscourse.com) and forms partly the basis of Chapter 3.
- With regard to leadership and management theory, Hans is greatly indebted to Ad Bossers and Joost van den Brekel at TU/e, who stimulated him to follow as many of their programs on management and leadership as he could. Hans also acknowledges two executive chairmen of TU/e, Henk de Wilt and Amandus Lundqvist, who, in different style, represented superb examples of authentic and visionary leadership in action.

https://doi.org/10.1515/9783110468892-203

Together, the authors express their sincere gratitude to the following:

- Yong-Wang Li, founding manager of Synfuels China Technology, Co. Ltd., who warmly embraced the idea of a SynCat Ac@demy with scientific leadership courses and provided generous financial support for the project.
- Jens Rostrup-Nielsen, for sharing his views on leading scientific research in industry and academia and for actively contributing to the SynCat courses. We thank him cordially for his friendship and mentorship on the topic of leadership in research environments. We are grateful for having his views in a guest column in this book.
- Graham Hutchings. He and Hans together presented courses for the c*change network in South Africa, before the idea for a book even existed. More recently, they also presented the course for the Cardiff Catalysis Institute at Cardiff University. Together with both authors, the three presented a course at the University of Cape Town in January 2017. We cordially thank Graham for his guest column in this book.
- Michael Bowker. Mike contributed actively to several courses, provided constructive feedback, and occasionally warned us when we took it too far ("beware of the glazing eyes in the audience…").
- Michael Claeys and Jannie Swarts, for organizing short courses and enthusiastically sharing their insights, at the Universities of Cape Town, and the Free State in Bloemfontein, respectively, in the beginning of 2017.
- Karin Sora of De Gruyter, Berlin, long-time friend in publishing Hans' books, for her warm support of this project.
- The ambitious staff of SynCat@Beijing, Synfuels China Technology Co., Ltd., and SynCat@DIFFER, Syngaschem BV, for their active participation, contributions, and constructive critical feedback on the course programs. It is a great privilege for us to have had the opportunity of interacting closely with so many talented young scientists – the scientific leaders of the future!

Hans Niemantsverdriet and Jan Karel Felderhof,
Nuenen and Eindhoven, the Netherlands, June 2017

# Special foreword

When Hans Niemantsverdriet and Jan Karel Felderhof asked me to write a few words of introduction to this book on scientific leadership, I read the manuscript with great interest. It is a great analysis that would have come in very useful in my own situation of many years ago, when I was struggling to start up my career in research. Had I read this book, I could have done my job of the past 20 years more efficiently.

Over the past years and before this book was written, the authors and I worked closely together in establishing the laboratories SynCat@Beijing and Syngaschem – SynCat@DIFFER. Jan Karel and Hans have done excellent work in helping our young postdoctoral fellows to start their work and bring it to a higher level.

Young scientific researchers are facing many challenges, not only to their own scientific qualities but also to adapt to the present competitive scientific climate; the latter is almost more influential for their careers. The funding system is one of the most critical factors. Young researchers have little other choice than to hunt for funds, and if they are successful, the grant is normally only small, say less than US$200 000. The odds are that they get discouraged after repeated disappointments and they leave science, implying they have wasted valuable time and we do not benefit from their potential. I personally have addressed this several times in lectures for executives and policy makers: If scientists are going to be totally controlled by the funding system, science will become a dead-end street!

Proper training of management skills and leadership makes the young scientists better prepared for facing and coping with these and other problems – not only how to compete for the scarce money but also how to train oneself to get along well with people and interact effectively with colleagues or the leaders above you and the students who work under you, and, importantly, also how to optimize one's performance in front of an audience and create beneficial publicity.

In my entire career, much of my working time, and some precious private time as well, has been spent on getting funding, on management affairs, and on the most difficult challenge, interacting with people. I would have loved to devote much more time to my interests in science.

I wish I had known Jan Karel and Hans much earlier!

June 2017
Yong-Wang Li
Professor, Chinese Academy of Sciences
Founding Manager, Synfuels China

https://doi.org/10.1515/9783110468892-204

# Contents

# 1 Creating success in your own scientific world

## 1.1 The world of research and development is highly competitive

Let's assume you are a young academic. You have been appointed as assistant or associate professor at a well-established university a few years ago or as a junior researcher in a research institute and you are working hard to become an acknowledged scientist in your research field. You have a dream about an important scientific challenge that you hope to solve in the coming years. You are definitely qualified for your job because you hold a PhD from a good university, you successfully finished a couple of postdoc positions at internationally renowned research institutions, you published together with famous scientists, you have several articles in high-impact journals, and you present talks at important conferences. Your peers are starting to know and respect you; you succeeded in getting one or two grants from your national science organization to fund a substantial part of your research. You have a few PhD students and a postdoc working for you, each on their own interesting project. Your teaching load may be substantial, but the students appreciate your enthusiastic lectures and some are eager to work for you in laboratory projects. In other words, it seems that you are well on your way in your scientific career.

Not everything is going well, however. You are working long hours every day and the balance between private and work life is different from what you, your partner, or your family would like. Daily issues in your work seem to determine your agenda and dominate your thoughts. Little irritations are sometimes affecting the atmosphere in your research group. Too often, you cannot find the time to talk to your students on a regular basis. Your head of department sometimes criticizes you for missing deadlines of important reports or other requests of information. The dean recently asked you to clarify what the focus of your research is, seen the variety of subjects you are covering. Also for you, it is not entirely clear what your group will be doing a few years from now. This is also caused by the diversity of funding sources that you have to deal with and the lack of consistent sustained funding for a particular project, maybe even the one you personally like most.

**Experiences of First-Year Assistant Professors**

A recent survey among 111 newly appointed assistant professors at mostly research-intensive (85 out of 111) American universities on their experiences after the first year revealed that 93% have worries about the current funding environment, while 81% feel pressure to present preliminary data or publish quickly.

As the most challenging aspects of the first year, the young professors mentioned managing time (34%), getting the laboratory going (23%), managing students and postdocs (15%), and teaching (10%).

https://doi.org/10.1515/9783110468892-001

Mentoring students was seen as the most rewarding task of their professorship by 52% of the respondents; 33% mentioned getting results in the laboratory, and 9%, teaching. Experienced faculty was by far the most important source of help and advice (62%); friends from graduate school, previous mentors, or others accounted for 12–13% each.

About 60% of the first-year professors felt that their work-life balance was less good than before, although 23% found that it had improved.

Source: New Professors by the Numbers, Chemical & Engineering News, CEN.ACS.ORG; American Chemical Society, May 22, 2017

This situation is very typical for young academics – and sometimes for the more senior professors as well. As academics, we learnt very well how to do our research and how to explain things to others, or even how to plan our research projects. However, insights in organization, strategic planning, managing, and how to be a leader are usually much less developed, as only a few academic organizations pay sufficient attention to it. Here are some of the key questions:

- How do you organize all your activities, and how do you maintain a proper balance between them?
- How do you judge if you are on the right track to becoming a tenured professor, an influential scientist, or a reputed educator in your field of science?
- Are you sufficiently clear about what you want to achieve with your team, and how do you ensure that your team's activities create synergy?
- How can you become a true scientific leader, so that you can make a difference with research that will have impact?
- And do you actually know how success is defined for your activities? Excellent papers in top journals? Prizes and Awards? Your own future research institute? Having delivered a number of successful graduates that are easily finding their way in society?

### 1.1.1 The academic climate has changed considerably over the past decades

Historically, universities were the exclusive seats of scientific research, which was done in close connection with teaching. A student was mentored by a professor, under whose personal supervision he studied his subjects and did his research, be it experimental or theoretical, in a close mentor-pupil type of relationship. Von Humboldt's ideal of academic freedom at independent universities was generally accepted; the role of the government was limited to providing funds and setting the legal framework for science and education. The chair holder or professor, once appointed, had full freedom to determine his research subjects. Teaching was mostly arranged within the professor-student mentorship and on a largely individual basis, which later changed to educational programs organized by the faculty, the joint academic staff of a university department. The implicit understanding was that academics are intrinsically motivated to strive for excellence, and the climate was to "leave science to the academics." Output, such as publications or dissertations, was not assessed by external parties in any way. Academic quality was merely how good the professor was in the eyes of his/her peers. An important factor was that access to academic education was limited to students who (or whose parents) could afford it, and in practice, the student/staff ratio was very much smaller than it is now. In terms of scientific quality, the academic landscape had plateaus, several valleys – some of these deep – and incidentally peaks, and insiders knew very well where to go for scientific excellence.

**Von Humboldt's Ideal of Higher Education**

Wilhelm von Humboldt (1767–1835), philosopher, minister, and diplomat in Prussia, was a strong believer in academic freedom for professors and their students and in the integration of education, arts, sciences, and research, to offer comprehensive studies along cultural knowledge in a single institution. Here, students would be allowed to choose their own course of study, enabling them to become autonomous individuals and world citizens by developing their own reasoning powers in an unbiased environment of academic freedom.

He founded the University of Berlin, which was later named after his two-year younger brother Alexander. Von Humboldt's concept was widely followed, first in North-Western Europe, and later also in the United States and other parts of the world.

In the 20th century, and notably after the Second World War, science and research gradually were no longer the exclusive domain of universities but also of governmental or private research institutes and not to forget the industry. Companies like IBM, EXXON-Mobil, Siemens, Philips, and Shell were known for their large corporate labs, which sometimes were deeply engaged in fundamental research and, as real powerhouses of scientific knowledge and development, almost functioned as an academic research institution.

In the academic world, universities saw the number of students rise, leading to much higher and less favorable student/staff ratios than before. Massive classes with several hundred freshmen in a lecture were no exception anymore. Funding did not increase proportionally, and scarcity of funds naturally led to the involvement of politics in decisions on how to distribute budgets over the various institutions. National funding agencies acting on behalf of the government started to play an increasingly larger role in the distribution of funds based on proposals, and hence, universities had to attract an increasingly growing part of their income through applications, of which only a fraction could be granted. Acquisition of funds from external sources in competition became the norm. To coordinate all this, the management of universities was professionalized to levels common in the business world. Management theories were introduced in which distribution of resources is based on quantitative parameters to measure performance and output (the so-called key-performance indicators), which are perceived to reflect the quality of research and teaching.

#### 1.1.1.1 Striving for excellence

Around the turn of the century, 2000, "excellence" became the key word in the global academic world. Many countries established "excellence" criteria and started to rank their universities accordingly, and if countries chose not to make such distinctions, then international organizations, magazines, or even universities themselves made up such rankings, based on criteria such as impact of publications, awards received, size of international cooperation network, etc. As American top universities generally dominate such rankings, these institutions were increasingly seen as ideal examples of the perfect academic institution, not only in Asia – where this was traditionally the case anyway in countries such as Singapore and China – but more and more also in the alma mater of the academic world, Europe.

In parallel, the public opinion (or perhaps more precise, the opinion of some politicians and influential captains of industry) became an important factor as well. Universities were criticized for being inefficient and overly expensive, allowing students to take much longer than the nominal time to complete their studies, and for doing research that was insufficiently useful for society. In particular, European countries felt that their ability to innovate was weak in comparison to, e.g., the United States or emerging economies in Asia. Universities encouraged their staff and students to become entrepreneurs, to start companies. Several universities, together with local government, founded innovation centers, science parks, and incubators to facilitate start-ups and interaction with all sorts of companies. The European Union (EU) created many programs in which international consortia of academic and industry and in particular small-medium enterprises could submit proposals for joint projects. The EU also started the European Institute of Technology, initially thought as a European MIT (modeled after the Massachusetts Institute of Technology in Boston) but later realized in the form of complex and largely virtual networks of Knowledge and Innovation Centres across Europe.

This is not to say that all of these trends are necessarily negative. Several successes of innovations starting from universities can be claimed. Also, insight in entrepreneurial activities widens the horizon of students and can prepare them better for their careers. However, universities should not forget what they are for, namely, education and training in science and technology at the highest possible level. In this sense, a famous book from the early 2000s appeared (Lewis 2005), entitled Excellence without a Soul – how a great university forgot education; in the language of the internet, university.edu became university.com.

Another trend in the 21st century is that national or regional governments started "excellence initiatives," where universities or even clusters of universities could compete for star status and be recognized as "even more excellent than the rest." In some countries like the Netherlands, a significant part of the direct funding for academia was taken away and added to competitive funds of the national science organization. Successful initiatives received substantial financial support, but all in all, the success rates of proposals for individual academics decreased to alarming levels of 10% and below, implying that scientists are forced to spend a considerable part of their time to write proposals that are very likely to be rejected or raise funding in different ways. As proposals need to be reviewed by scientific peers, the pressure on the scientific world to review and assess proposals and participate in juries also increased considerably. The imbalance between highly rated proposals and available funds meant that many high-quality initiatives did not receive the means to realize them.

#### 1.1.1.2 Publications, awards, and other academic currency

Excellence needs to be demonstrated. Publications are probably the most important academic currency in this respect, and their number is growing exponentially. The growth rate of cited scientific publications has risen from less than 1% before the middle of the 18th century to 2–3% in the first half of the 20th century and 8–9% today, implying a doubling of publications in less than every 10 years (Bornmann & Muetz 2015). This is partly caused not only by the advent of research in emerging countries but also because of the "publish or perish" climate felt by academics all over the world, and not to forget the fact that many universities nowadays request students to publish at least one or more papers in established journals before they can submit their PhD thesis. Of course, all these submissions have to be handled by editors, reviewed by peers, and added to the activities of the academics. Who has time to read all these articles? Do we have to?

Prizes and awards form a second indicator of successful scientific performance. No one will question that a Nobel Prize winner is an exceptionally good scientist. There are undoubtedly other awards with very rigorous and objective selection criteria. However, over the last decades, we have also seen the advent of a flood of prizes, awards, or recognitions for which the selection procedures are doubtful and dominated by political considerations. Also, professional societies, with the aim to offer opportunities to their members to distinguish themselves, have installed many so-called awards, which in essence are no more than modest stipends for visiting a conference or spending a few weeks on sabbatical leave. Universities have installed prestigious prizes to honor famous persons (not necessarily always scientists) just to benefit from the publicity that such an event generates. It is not at all uncommon to find scientists who hire professional assistance to propose themselves or their students for recognitions or use their personal influence to get invited for prestigious invited lectures at leading conferences.

#### 1.1.1.3 How do academic enterprises define success?

Anno 2017, when we wrote this book, many academic institutions have developed into enterprises largely led by managers rather than by intrinsically motivated academics, and all these universities are in competition for the best students,

postdocs, academic staff, and funding. According to Binswanger (2014), the world of scientific research is dominated now by what he calls “artificially staged competitions” on many levels:
- Universities in national and international rankings of many sorts all do their best to prove that they are “excellent” (for whatever that means).
- Departments among each other within their university for resources or outside in their respective disciplines for scientific status.
- Research groups in or outside their universities and institutes, on the basis of publications in high-impact journals, citation scores, personal awards, invited lectures, or in nationally organized peer reviews, where each group receives scores on a scale of 1–5, for example.
- Students in their departments, for “Best Student of the Year Contests” and alike.

This the climate in which you will have to find your way, whether you are a young assistant professor working hard to get tenure (a so-called “tenure tracker”), the newly appointed head of department, or an established scientist who has just been appointed as the new director of a research institute. You will have to live with the pressure of the race for excellence, but it is important to once in a while reflect on the priorities that really matter and ask yourself the question what success means for you (see Tab. 1.1). Of course, favorable indicators will result if your research group is successful in terms of the true values in the right column of the table. However, we see too many young scientists who are entirely focused on the short-term successes of good performance indicators, assuming that these are the road toward success in the long run. Understandably, they copy what they have seen from supervisors with similar philosophies, but nevertheless questionable as a pathway to sustained success.

Tab. 1.1: How do you define success (for the institution, yourself, your funding organization)?

| Key performance indicators (KPIs) | True values |
|---|---|
| Number of scientific publications? | Quality of your graduates? |
| Papers in Science or Nature? | Contribution to ‘knowledge’? |
| H-index? | Recognition for quality of teaching? |
| Large-sized research group? | Well-equipped laboratory supported by one or more experienced technicians? |
| Number of prizes/awards? | Recognition for the high quality of a coherent research program? |
| Number of graduates from your laboratory? | Success of your alumni’s careers? |
| Amount of funding or number of grants acquired? | |
| National or International Ranking? | |

## 1.2 Trends in governance, control, management, leadership

Scientists have always had an intuitive resistance against being told what to do. It is in their nature to discover, to go for the adventure of the unknown. It is almost impossible to plan for discovery, and hence, the typical project management idiom of Gantt charts, deliverables, milestones, or the complicated collaborative schemes that are asked so often nowadays in funding programs do not appeal to the average scientist. Nevertheless, they have to play the game to survive in the present academic climate, and many found ways to cope with such systems and still keep capacity for true "blue-sky-discovery-type" research activities. At the same time, academic institutions, research organizations, and funding bodies adapted increasingly the governance and control principles from the entrepreneurial business world, as we sketched above. These principles themselves are far from static either, and it useful to sketch briefly how these changed over the past century.

### 1.2.1 Historic developments in management, governance, and control

Managing a production factory is something else than directing a bank, coaching a sports team, or leading a university. Different situations require different approaches to governance and control, as the following examples illustrate.

- Production calls for a task-oriented approach; think for example about the way Henry Ford used the division-of-labor principle in the assembly line for producing the T-Ford in the beginning of the previous century. The management concept is actually based on hierarchy in activities and tasks, according to the theory of Frederick Winslow Taylor (1911), an engineer also known for his invention of carbon steel tools.
- Administrative processes need a procedure-oriented approach, with pure bureaucracy as the ultimate form. In the early 20th century, the German sociologist Max Weber saw bureaucracy as the most efficient and objective way to organize processes and hierarchy, with the benefits of maintaining order, optimizing efficiency, and avoiding favoritism.
- Science, as well as the domain of professional services, requires a knowledge/know-how-oriented approach, as in Maister's description of the management of professionals (Maister 1993). Weggeman (1992) actually advocates that professionals flourish most when subjected to a minimum of management – you can imagine that he is one of our favorite leadership gurus ☺.
- Sales and project management were considered to best require a result-oriented approach, which lead to the principle of management by objectives or management by results, which we owe to Drucker (1954).

All these approaches were quite focused on their own domains of application, and they did not always fully result in the desired outcomes. A basic shortcoming of the early theories was that people and human interactions did not feature explicitly. This started to become evident in the middle of the previous century. We cite a famous article by Tannenbaum and Schmidt (1958) from 1958, in which they described the continuum between fully autocratic, boss-centered leadership to the fully democratic management style in which task and responsibilities are delegated to the staff see Fig. 1.1. In essence, they explicitly acknowledged the value of teamwork. A few years later, Blake and Mouton (1964) proposed their famous managerial grid, an early assessment tool for management style based on two parameters: task orientation (performance) and attention for good human relationships. People management was established. However, leadership entails more than management of tasks and people. We discuss this at length in the coming chapters of this book.

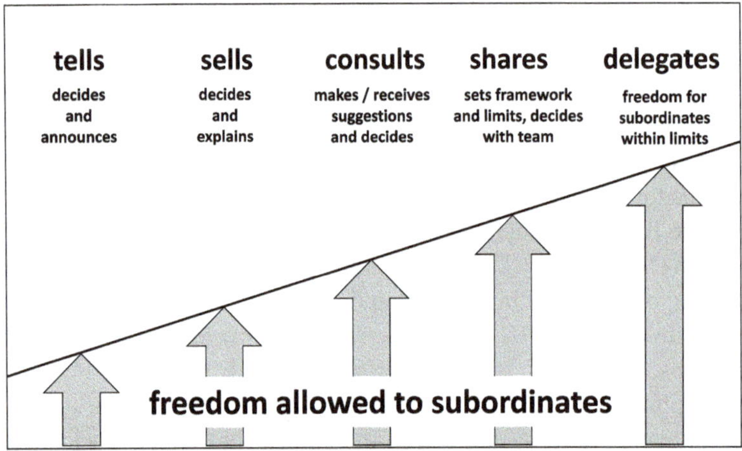


Fig. 1.1: Management style versus room for initiative and spontaneous action left for employees.

On the institutional level, governance and control theories initially concerned pure task management and control over results, while the state of the financial budget dominated the thinking. In the second generation, strategic planning according to the theories of Mintzberg (1978) and Porter (1980) played and increasingly important role. Governance and control interest focused monitoring the organization's progress in realizing its strategic plans and on the factors that determine financial outcome. The Balanced Scorecard was developed by Kaplan and Norton (1996) to keep track of success factors and key performance indicators. We described above how this thinking and methodology became common practice in the academic world. In essence, this second generation governance and control is a top-down mechanism. It leaves little freedom for initiative and spontaneous creativity.

At present, we see in modern organizations, for example, in many start-ups, the transition to a third generation (also referred to as authentic generation) of governance and control, which emphasizes the importance of comprehensive insight in the complexities that play a role (Felderhof 2007). Flexible organizations require the intelligence and responsibility of people on several levels for making decisions on how to act in specific situations. For this, people have to understand the goals and the needs of their organization thoroughly, and they need to dispose over sufficient freedom to operate and to decide. Terms like “self-steering,” “self-management,” and even authentic “self- leadership” have entered the scene; hierarchical structures tend to become less dominant. In its ultimate form, third-generation governance and control rely on the unique abilities and strong points of individuals, who together operate like birds in a flock (see the cover illustration). The situation determines which abilities are needed most, and as in an effective flock of geese, the leadership changes accordingly in an almost automatic fashion.

Is it necessary to know all this if you want to become a successful researcher in a university, governmental, or industrial institute? We think having insight in the playing field and the rules of the game, whether these have been spelled out or are largely unwritten, and recognizing how institutions, be it your department, your university, the academy of sciences, or the company to which your institute belongs, are gradually changing their organization structures, will help you in finding the way to achieve your goals.

### 1.2.2 The scientific leader of the early 21st century

We will define what we mean by leadership later on in Chapter 2, but for the moment, we assume that you are inspired by some brilliant idea, a greater plan, a dream, a vision of what you would like to achieve over time. Maybe it is solving a compelling problem, developing a new technology, sharing your enthusiasm for science with young students, or reaching the (perceived?) status of a successful scientist, that appeals to you. No matter if you only have it vaguely in your head or clearly spelled out in a few power point slides, there will be an inner driver behind where you are now.

Science is teamwork; very few scientists work alone. You will have students, postdocs, and perhaps assistants who work with you, and perhaps, you will have collaborations with your colleagues. Your students will soon have been born after 2000. They belong to the so-called “Alpha generation,” have probably been educated in a different world than you, and may have a different approach to learning; see Tab. 1.2, which presents some trends, differences, and expectations (Lemmens 2017).

Tab. 1.2: Characteristics of generations from the 20th and 21st centuries and skills anticipated to become essential in the near future.

| **20th century generations** ***"Baby boomers, Gen X, Y, Z"*** | **21st century generation** ***"Generation α"*** | **Requirements for graduates of 2030** |
|---|---|---|
| *Verbal* | *Visual* | |
| Sit and listen | Try and see | Systems thinking |
| Teacher | Facilitator | Integrative and connecting |
| Job security | Flexibility | Transdisciplinary approach |
| Commanding | Collaborating | Creative |
| Curriculum centered education | Learner centered education | Entrepreneurial |
| Closed book exams | Open book world | Analysis *and* synthesis |
| Books & paper | Screens & devices | Global, mobile, agile |

The key toward building up a successful research group is how to lead and inspire your students and coworkers, whatever the differences in approach and style exist, in a way such that they get the most out of themselves, while at the same time they contribute optimally to your goals. That sounds obvious and is easily said, but how do you accomplish this?

**My legacy in science are my graduates!**

Asked what his legacy in science will be, 2017 Nobel Prize Winner Prof Fraser Stoddard answered: "My legacy will not necessarily be my chemistry, which has been described in more than 1,000 publications – too many! It will be the more than 400 graduate students and postdoctoral fellows whom I have trained and mentored. Almost 100 of them have gone on to be professors in their own right in universities all around the world, while many more have gone into industry, government, finance and publishing. It will be the young people I have trained who will be my legacy, particularly if they listen to my plea to tackle a big problem in science and not continue doing "Stoddard chemistry."

From the interview by J.-F. Tremblay, Chemical & Engineering News, March 13, 2017, pp 34–35.

As we describe in detail hereafter, scientific leadership rests on three pillars: a compelling vision of what you want to achieve, relational skills to inspire your team and to interact successfully with them, and managerial skills to organize everything that is needed to accomplish this.

In the early phase of your career, it is absolutely important to invest in the skills that help scientific leaders to accomplish their job on a daily basis. Several of these are related to academic practices, and we will have learnt these during our study time, but there are others that need your attention as well, for example, how to efficiently organize your work, manage your projects, teach your classes well, write articles and grant applications, and present with impact at conferences. Some chapters deal with these skills, and in particular, we stress the importance of presenting your work and ideas effectively in written and spoken form, by always having a clear message that will be well understood by the particular audience or readership that needs to understand the message and hopefully respond to it as you hope. These skills are commonly referred to as the hard skills, or the typical managerial skills. Mastering these is essential, but these skills alone are not a guarantee for successful leadership.

#### 1.2.2.1 Academic managers or academic leaders?

It is important to appreciate the difference between leadership and management: Managers deal with tasks, keeping the business running, and keep projects on track and often with a short- to medium-term focus. Leaders, on the other hand, focus on goals and the people who work hard to achieve these, with a long-term perspective. Hence, leadership includes all the management roles needed to keep the team on track, but it is definitely much broader than that. The comparison in Tab. 1.3 clearly catches the difference.

Tab. 1.3: Differences between management and leadership.

| Managers | Leaders |
|---|---|
| Handle tasks | Motivate and inspire |
| Analyze data | Identify patterns and trends |
| Weigh alternatives | Handle the unpredictable |
| Solve problems | Judge how to cope with uncertainty |
| Take decisions | Support others to take decisions |
| Have often short-term focus | Have a long-term vision |

#### 1.2.2.2 Relational skills, authenticity, and charisma

Some professors, group leaders, or directors seem to have been born as natural leaders; they give us the impression that they feel confident and at ease in everything they do. The question is if they feel it themselves also. Nevertheless, it is probably very true that experience and insight acquired over the years, in a mix of successes and failures, contribute significantly to self-confidence and the ability to act in an authentic way.

Authenticity and charisma are probably the most emphasized factors for successful leadership. These, however, are not characteristics that are easily acquired by reading books or following a miraculous course. These are characteristics that people possess to some extent and have learnt to develop further and rely on. You will probably feel best and be most successful in your work if you can behave according to your own genuine personality, your "authentic self." It implies that you feel at ease, you know your strengths and your weaknesses, and you communicate with others respectfully, in a relaxed and self-confident manner. Your self-confidence derives from knowledge and know-how, managerial and organizational skills, and insight in processes and human relations, which you achieved yourself or learnt from supervisors and mentors, combined with self-reflection and self-coaching.

Charisma is another often quoted attribute of "natural leaders." The term is not easy to define and is used in a variety of meanings, but let us, for simplicity, call it the ability to inspire others on the basis of warm personal relations. The word charisma comes from the Greek language and can be loosely translated as "giving in a graceful manner." Charismatic leaders have a natural talent for making others enthusiastic for their views and goals that often derive from a compelling vision. Can one learn how to become charismatic? We don't think so. You can learn to behave authentically, and if you as a leader believe in the power of synergy in your interaction with students, coworkers, and colleagues on the basis of a common ideal, where all can benefit in terms of their own goals as well, then chances are good that others may start to see you as charismatic. In other words, charisma is something that hopefully will come and start to be recognizable at some stage.

As the leader, you want to be the example for your students, your coworkers, maybe even your colleagues. Is there a secret recipe for becoming a scientific leader who inspires, educates, leads his/her team toward the realization of his dreams, and enables the team members to accomplish their goals? If there is a secret, it boils down to creating the environment and conditions under which your students, postdocs, and coworkers and yourself can flourish to achieve the best for themselves and simultaneously for you and your organization.

Our philosophy is that you can only be a true and inspiring leader if you have first learned how you can lead yourself, i.e., that you feel confidently at ease in your role of heading a group. As an accomplished and authentic self-leader, you will automatically inspire others and ensure that the people working with you progress through the same development and achieve their own status of authentic self-leadership. Is this a magic recipe? Not at all, we think it is mostly applying healthy portion of common sense....

## References

Binswanger, M.: Excellence by Nonsense: The Competition for Publications in Modern Science. In: Opening Science; Bartling, S., Friesike S. eds. Springer, Berlin; 2014; pp: 49–72. DOI: 10.1007/978-3-319-00026-8_3.

Blake, R., Mouton, J.: The Managerial Grid: The Key to Leadership Excellence; Gulf Publishing, Houston, 1964.

Bornmann, L., Muetz, R.: Growth rates of modern science: A bibliometric analysis based on the number of publications and cited references. J Assoc Inf Sci Technol; 2015; 66; 2215–2222.

Drucker, P.: The Practice of Management; Harper, New York, 1954.

Felderhof, J.P.K.: Enhancing Servant Leadership through Inspiring, Enabling and Focussing Enhancing Self-leadership; Informance Publishers, Eindhoven, 2007.

Kaplan, R.S., Norton, D.P.: The Balanced Score Card, Translating Strategy into Action; Harvard Business School Press, Boston, 1996.

Lemmens, A.M.C.: Eindhoven University of Technology, private communication, 2017.

Lewis, H.R.: Excellence without a Soul: How a Great University Forgot Education; PublicAffairs, Boston, 2005.

Maister, D.H.: Managing the Professional Service Firm; Simon & Schuster, New York, 1993.

Mintzberg, H.: The Structuring of Organisations; Pearson Education, London, 1978.

Porter, M.E.: Competitive Strategy: Techniques for Analyzing Industries and Competitors; The Free Press, New York, 1980.

Tannenbaum, A.S., Schmitt, W.H.: How to choose a leadership pattern. Harv Bus Rev; 1958; 36; 95–101.

Taylor, F.W.: Principles of Scientific Management, Harper, New York, 1911.

Weggeman, M.: Leiding geven aan professionals (in Dutch); Kluwer, Dordrecht, 1992.

Jens Rostrup-Nielsen

# Guest Column: Research should be managed by motivation – not by control

*Dr. Rostrup-Nielsen is former research director of Haldor Topsøe A/S in Denmark and is a founding member of the European Research Council. In this guest column, he gives his view on how to manage responsive research in an industrial environment* (Rostrup-Nielsen 2016).

Haldor Topsøe has always believed in a multiple approach to industrial research. It was his firm belief that understanding the science behind the development work is of crucial importance. First, this gives a strong basis for coping with problems, and it has great value for marketing activities too when potential customers notice that a company has deep and comprehensive understanding of the technology they offer. Much of the basic research can be published, which motivates the scientists and forms a basis for valuable collaborations and joint publications with university groups. Such collaborations can be sustainable only when they are two-way processes, where both sides benefit. Hence, it is essential that a company is active in basic research as well.

A second strategic component in Haldor Topsøe's philosophy was to be always at the forefront with core competencies, from advanced computational modeling to the newest methods for materials research, e.g., state-of-the-art electron microscopy and the use of big research infrastructure at synchrotrons. Being active in the further development of key research methodology yields competitive advantages for doing high-profile research. It makes the company an attractive partner for leading academic groups, and it creates the ability to catch opportunities for new directions in contact with the best of the field.

The third ingredient in the multiple approach was exploration aimed at "radical innovation", working on solutions for the next generation of technologies. To some extent, this can be done in-house or by outsourcing test attempts in the academic world and of course by actively monitoring what competitors are up to.

Managing all these parallel activities can be a challenge. In some way, handling the worlds of logically planned R&D and creative basic research is balancing order and chaos. I sometimes compare the management of innovative research with gardening. The meticulously laid-out baroque garden looks beautiful, but there are no surprises. A garden without any control, on the other hand, turns into an interesting wilderness with lots of activity, but probably not in the direction you want. Obviously, the truth is in between.

For the leader, the art is to manage with a loose line and to ensure that each scientist has room for one's own initiative. One can institutionalize this, for example, by using the 65-25-10 approach (which by the way I picked up from a research director at Shell): about 65% of the resources in a project are used for development work with well-defined milestones and budgets, and 25% is used by the team for its own solutions to the problem. The final 10% is available for any activity that a researcher finds useful within the scope of the company, sometimes referred to as the "Friday Afternoon Experiment". In this mode, the scientists have a lot of opportunities to contribute their own insights and have their own initiatives. Strict planning of activities is inevitable for large development projects, but it is a killer for projects in the explorative phase. Research should be managed by motivation, not by control.

Hence, for me, managing responsive research is striking the balance between "focus and hocus pocus", avoiding that innovation processes become too bureaucratic and ensuring that each scientist has room for one's own initiatives.

## Reference

A longer version appeared as Rostrup-Nielsen, J.R.: 50 Years in catalysis. Lessons learned; Catal Today; 2016; 272; 2–5.

# 2 Leading yourself and others in research

## 2.1 Introduction

In this chapter, we first analyze the sort of tasks that a researcher with a team of students and coworkers has to deal with, and we approach it from the viewpoint of the researcher self. We acknowledge that each person is different and that a standard personality does not exist at all, while also the environments in which people work may vary widely. We adopt a simple four-dimensional model that has proven its value in describing the way different people act in their work, and we apologize at the onset for the severe oversimplification that we hereby introduce. With two self-tests, you can score yourself on your operational personality dimensions, as well as on your approach to daily duties as expressed in the dimensions of management. We end with a practical definition of leadership but stress that self-leadership is a necessary requirement for becoming a true leader.

## 2.2 Your tasks as a research leader

Let us try to make an inventory of the various tasks that academics have to accomplish these days. Many of these will be important for research leaders in institutes or industry as well, perhaps with the exception of teaching (although new team members need education, instruction, and encouragement to dive into the basics of their new activities as well). Often, universities assess their academic staff on three major aspects, namely, education, research, and organization, while sometimes contributions to exploitation of research, innovation, and/or commercialization are considered in addition. We attempt to make a list (and anticipate that it may on one hand not be complete, while on the other not all these tasks may apply to you):

- Education: preparing and teaching classes, instructions, preparation of syllabi, handouts, exercises, question hours, preparing exam questions, correcting exercises, draft theses, grading exams, devising experimental practice courses along with written instructions, supervising and mentoring students
- Research: generating ideas and formulating research questions, reading literature, designing experiments, doing the actual research either by yourself or together with students, analyzing and interpreting results, writing reports and publications, going to conferences, preparing presentations, supervising research students and team members
- Organization: obtaining funding for research, doing administrative duties, planning, reporting, arranging group meetings and seminars, sitting on committees, networking, taking care of public relations

https://doi.org/10.1515/9783110468892-002

- Exploitation (although perhaps limited to specific universities only): pursuing commercial exploitation of research, handling intellectual property issues (patents, etc.), devising a business plan for the initiation of a start-up

All in all, this is an impressive list of activities, although not all may apply to you. Note also that many of the tasks in the list are typical management activities, also in the categories that are typically based on "science" or more generally "content" and "knowledge," namely, teaching and research. Mastering skills as time management, project management, strategic planning, and relation management, together with "soft" skills to effectively supervise, coach and motivate people, and communicate with them through spoken and written word, is as important as the academically acquired scientific and educational skills. How do you cope with all these challenges and how do you ensure that each aspect or each task receives the attention it deserves? Everybody is different and has his own way in dealing with the complexities of life in academia or, more general, research and development. We think the following analysis helps in appreciating how different people deal with the complex tasks of managing their work.

## 2.3 Different ways in which people approach their work

We almost enter the field of psychology when we try to describe the ways and attitudes in which people approach their work and all the activities that belong to it. In Chapter 6, we go a bit deeper into the subject, but here, we will assume that people's behavior in work situations can be described by the four simple operational personality dimensions: feel, think, do, drive (to be understood in the sense of achieve or perform).

Figure 2.1 shows the four personality dimensions that we believe describe the way people operate in their work: We all have the same blueprint in us for these

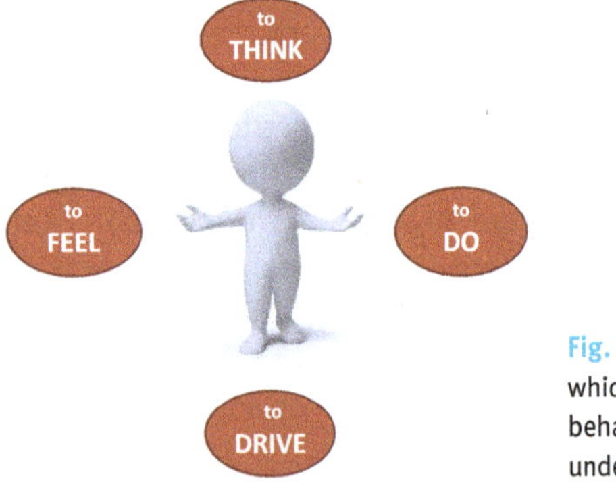


Fig. 2.1: The four operational personality dimensions, which we use as a simplified model for describing our behavior in the work place. To drive is meant to be understood as to the power of will to achieve.

dimensions and we developed them to variable extents. We describe them in the context of an academic research environment.

The *Feeler* in us is empathic, sympathetic, and sensitive and reacts on the basis of intuition and first impressions. Feelings connect us with the subtleties of intangible intelligence. Emotions regard the positive (winning, success) or negative (loosing, frustration) reactions to activities or outbursts of suppressed feelings. Intuition gives us immediate directions on what to do in an imminent situation. Our feeling tells us whether such response is ethical and sustainable in the long-term.

The Feeler is strongly interested in personal interactions with others and is highly sensitive to emotions that first encounters with situations generate. He/she strongly appreciates the value of networking, of good relations, and of fruitful contacts with peers in the organization and in the scientific or professional community. It is sometimes said that first impressions are determined by human emotions and feelings, before the ratio takes over.

The *Thinker* in us likes all content-related aspects of the work, e.g., the intrinsic value of the research, the design or interpretation of new experiments, or what the literature has to say about a particular scientific question. This type of person may also like to think strategically about how meaningful activities are in a larger context or, for example, how the research and educational efforts can make a difference and will have impact. The Thinker likes to carefully plan activities before they are done, see the cartoon in Fig. 2.2.

Fig. 2.2: While the Thinker is still thinking what needs to be done, the Doer is already doing something (reproduced with permission by Dan Reynolds).

The *Doer* thrives on being active, be it in the laboratory to do experiments or in the execution of tasks in general. If there is a plan that the Doer believes in, he/she likes to carry it out. He/she usually values structure and organization, such as clear plans and protocols for experiments and standardized templates for reporting the results, as these provide clear and generally accepted frameworks for how things should be done. The Doer is usually very active in accomplishing targets, networking, and professional activities such as education.

The *Driver* (drive to be understood as the power of will to achieve) has focus on output and tangible results and gets energy from completing tasks and satisfaction from the fact that a deadline has been met or a report or paper has been submitted in time.

**Simplified Picture of the Four Management Dimensions in a Company**

Allow us to sketch a stereotypical (and in fact too simple) division of roles in a company that designs and produces consumer goods. In the R&D department, the researchers are creative and strong in the think dimension. The developers are more focused on the technology and add the drive dimension (for in the end, it has to work). Engineers add the do dimension for the product that has to be tested and eventually produced. Designers add the feel dimension, for clients will have to like the look, feel, and use of the product. Business developers and sales employees are again focused on the drive dimension, to sell the product and make a profit.

### 2.3.1 Balance

Although each of us possesses these characteristics by nature in a certain mix, it is important that we are aware of our preferences and learn to compensate at least a bit for the ones that are too little present.

If we score ourselves on these four personality dimensions, we obtain graphs like the ones in Fig. 2.3. The balanced region is somewhere in the middle of each axis.

What do these graphs tell us? The person represented on the left of Fig. 2.3 thrives on doing things and has a strong drive to get things done. His scores on the feel and think dimensions are significantly lower. We could imagine that this profile corresponds to a manager who has a clearly described set of tasks, or perhaps a production worker in a factory, or an analyst in a medical laboratory. The person on the right has a profile that looks more like a scientist, e.g., a professor with a large group of students working with him. He thinks a lot, clearly likes to interact with other people (the feel dimension), and has the inner drive to make sure that things will get done, but the students will have to do it mostly. We would call this a "balanced profile" because all dimensions are reasonably to strongly developed, without one being really too dominant.

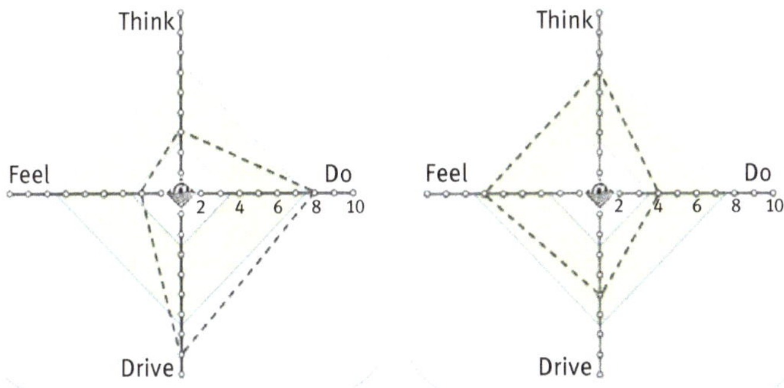


Fig. 2.3: Two examples of personality dimension scores. Left, a typical performer who will ensure that the organization produces predictable output and required results, and right, a more balanced profile, which could well be that of a successful professor who is fond of content and meaning and likes to interact with his people but at the same time feels a strong inner drive to realize his goals.

### 2.3.2 Knowing yourself

Knowing yourself and how you score on your personality dimensions is an important first step toward growth in your job. The Appendix has a scorecard for analyzing yourself and may be a starting point for reflection. Are your personality dimensions in accordance with the sort of position you have in mind? If you feel they are unbalanced, can you do something about it? You can also use the scheme to assess the people working with you. In a later chapter about human interactions, we will explain how this scheme may help you to recognize patterns in how people operate in teams.

There are of course many more important factors in understanding the way you work; we take Drucker's popular treatment on self-management as a guideline here (Drucker 2008). For example, recognizing the way in which you learn and deal with all kinds of information is an essential point to know. Do you mostly learn from reading texts, or by listening, or perhaps by explaining things to others? Some people learn by taking extensive notes of every presentation they hear and write notebooks full during conferences and seminars because this is the way they absorb, process, and remember information (sometimes even without ever looking back to their notes). Our school systems are largely based on traditional ways in which a teacher explains new material and students process the new knowledge by listening, taking notes, studying books, and making exercises. However, many people learn much more effectively in other ways, by doing, by trying out themselves, by searching the relevant information they need in the context of what they are busy with at that moment, and not because the professor or teacher tells it in a lecture that is delivered at a time that the student is perhaps not at his/her most perceptive state of mind anyways. So it is vitally important that you know your own most effective ways to learn and those of your coworkers and students.

Other important questions about yourself are the following:

- Do you produce the best outcome by working alone, or together with others? You can still be a very successful leader if you prefer to do some essential tasks entirely by yourself, but it is important that you realize this and your team understands it.
- Are you good at taking decisions or by advising others on key matters? According to Drucker, the promotion of a successful second deputy to the director position often fails because an excellent advisor is not necessarily an effective decision maker. Having thought about this is obviously important before you decide to go for the highest leadership position.
- When do you perform at your best? Under conditions of challenge and stress, or do you need a more comfortable, predictable situation? Do you feel at ease in a large and complex organization, or rather in a smaller setting?
- It is an illusion to believe that you can change yourself on such fundamental aspects of your personality. Starting from your strong points and continuously working on enhancing your strengths and acquiring additional skills is a good strategy to build your career on. This will then be the basis for your (self) leadership.

People operating from a strong inner drive may be interested in taking a deeper look at themselves as the basis for growth. Louise Hay sold more than 30 million copies of her book on the deeper patterns underneath a personality and how you can change them (Hay 1984). The authors have developed tests for giving insight in your self-leadership potential.[1]

## 2.4 People's operational management dimensions

The operational personality dimensions Feel, Think, Do, and Drive match well onto dimensions one encounters in the daily activities at work: Interacting with People, Content, Structure, and Performance, respectively (see Fig. 2.4).

The Feeler is generally interested in other people and likes to interact with them; the thinker likes content and knowledge. Doers are helped when it is clear what needs to be done, as described in procedures and protocols, which all relate to clearly structured work activities. Self-propelling doers like being active with their profession, reaching targets, or networking. The driver likes to generate output and have impact; i.e., he/she is focused on achievements.

We can now add all the tasks of the academic to Fig. 2.4 and thus obtain Fig. 2.6. We immediately see a relation between the type of activities various people will feel attracted to or maybe will be not so good at.

1 www.scientificleaders.com

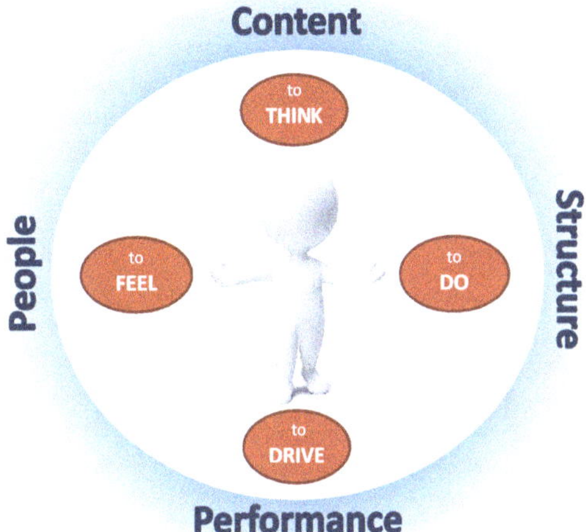


Fig. 2.4: Personal operational dimensions Feel, Think, Do, and Drive correspond to management dimensions People, Content, Structure, and Performance (getting results). A person functioning well in his/her relevant management dimensions is in control of realizing a good performance; that is, results and output are in balance with content and relationships.

You can assess your own dimensions with the score cards of Fig. 2.5, which are also available in the Appendix.

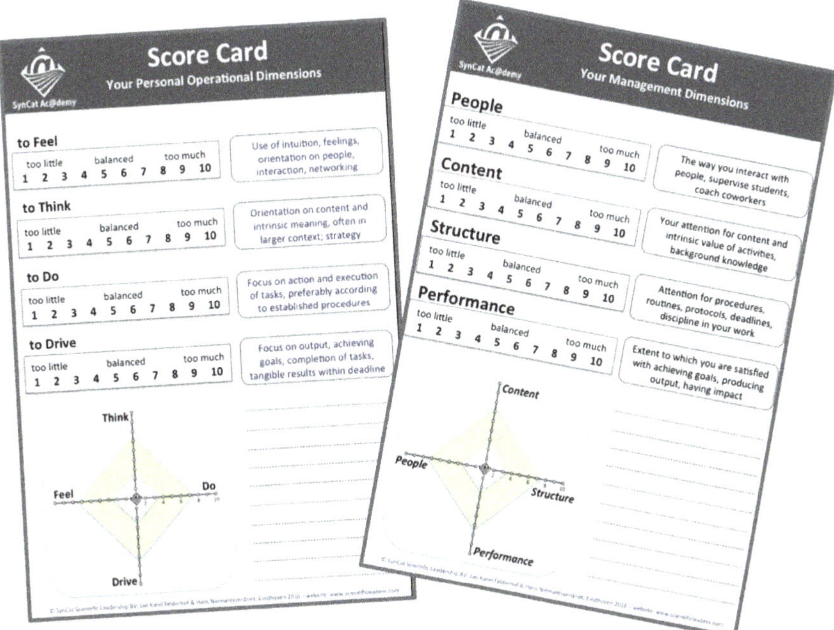


Fig. 2.5: The Appendix has scorecards for your personal operational and management dimensions.

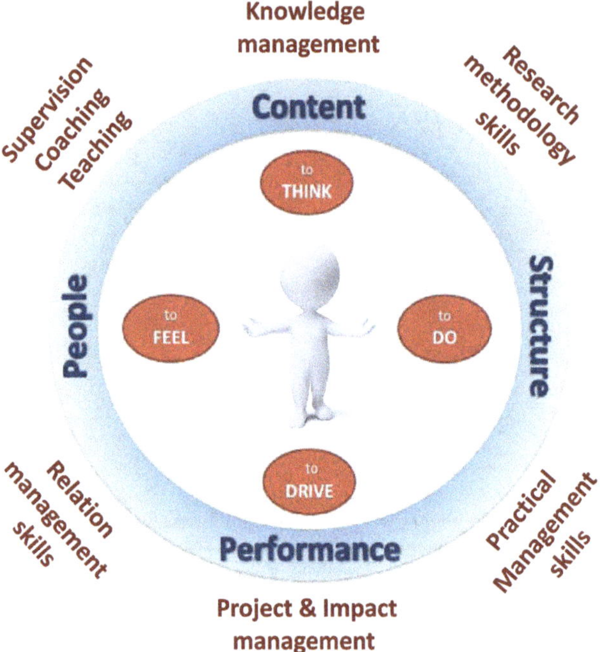


Fig. 2.6: Academics have to find a feasible balance among all the tasks that leading a research group entails. The person in the middle is you, the young academic, surrounded by all the activities and tasks you are involved in. The four dimensions Think, Do, Drive, and Feel represent important aspects of your personality, which determine and energize the way you operate.

## 2.5 Self-leadership

Now, what is an accomplished self-leader? It is somebody who has a good view on all what needs to be done, feels responsible for it, understands the relation between character and personal inclinations for types of work, and has achieved a way to balance his/her operational personality dimensions with the tasks that need to be accomplished. How? By learning the skills that he/she does not possess naturally and by tempering his natural inclinations to go for those aspects of the work that he/she feels most attracted to. It is a matter of learning the necessary skills, exploiting talents, relying on experience, and applying a mix of common sense, patience, managing your own emotions, and finding the right balance between all these factors. Above all, it is taking the responsibility for your group in your own hand and feeling the inner drive to take action where it is needed. This is *your* research group, department, and institute, and that comes with an obligation; you cannot hide behind others.

Blanchard and coworkers formulated the responsibility part of self-leadership in a very compelling way (Blanchard et al. 2005):

> *"Organizations need people who act as if they own the place."*

These are people who dare to take initiative if they believe it is needed and who feel personally responsible. They are empowered people who like to learn and develop further. They like to solve problems and they understand what is needed. And, very importantly, they are honest, sincere team players who have the confidence to be themselves and to behave authentically. This in contrast to people with a "victim mindset" who lack the courage to challenge what superiors say or to break through the constraints of "this is how things are done and have always been done in this organization." They often assume that authority or seniority is needed to raise critical questions or to propose novel or alternative ideas.

The following quotation from Jack Welch, former chief executive officer (CEO) of General Electric, nicely illustrates the difference between victims and self-leaders:

> *"You can look at the situation and feel victimized. Or you can look at it and be excited about conquering the challenges and opportunities it presents."*

Note also the subtle difference in how to involve someone's help: Do you ask "I have a problem, can you please help me" or do you say "I need someone who can..." The latter implies that you know what is needed, that you have a plan, while the former sounds rather helpless.

Does self-leadership derive from power, or from a higher position in the hierarchy? Not necessarily. It is in fact one of most frequently assumed constraints to believe that one needs legitimized authority to ask others to assist, or in other words to occupy a higher hierarchical position in the organization or have specific mandate from a superior

Tab. 2.1: Summary of self-leadership, after Blanchard and coworkers.

<table>
<tr><th colspan="2">Self-leaders take responsibility</th></tr>
<tr><td colspan="2">Organizations need people who "act as if they own the place"<br>– People that take initiative and feel responsible<br>– Empowered people who like to learn and develop<br>– Problem solvers who understand the needs<br>– Honest team-players who are their "authentic selves"</td></tr>
<tr><td colspan="2">Self-leaders challenge assumed constraints</td></tr>
<tr><td>"Assumed constraint": We have always done it this way...a belief based on past experience that limits you in your current situation.</td><td>Be constructively critical and open minded about ways to achieve goals.</td></tr>
</table>

Tab. 2.1 (continued)

| Self-leaders celebrate their points of power | |
|---|---|
| "Assumed constraint": Power derives from function in the organization: "I'm not in a position to get people to do what I want them to do..." | Recognize that there are various "sources of power": knowledge, relationships, personality, position, reputation, etc. |
| **Self-leaders collaborate for success** | |
| Help me to solve my problem, please... | "I need someone who can..." implies a plan and shows initiative |

to involve others. Power in a generalized sense also derives from having specific knowledge, experience, skills, relationships, personality, previous successes, and so on.

Being an accomplished self-leader implies that within your span of control, you take maximal responsibility and initiative to ensure that things will go the way you think it is best for your team, your laboratory, your organization. Table 2.1, adapted from the work of Blanchard and coworkers, (Blanchard et al. 2005) summarizes the most salient characteristics of self-leadership.

## 2.6 Toward scientific leadership – a practical definition

We return to Fig. 2.6. When you succeed in giving all the tasks of a scientist the needed attention in a way that works for you, it is time to consider how you are leading your group. If you are a relatively young assistant professor, your group will most likely consist of a few PhD students and maybe a postdoc. In the future, you may grow to a full professor with a large group of assistant professors, postdocs, and students and, if you are so lucky to have them in your team, even some technicians and an administrative assistant. No matter if you are the novice or the experienced head of department, successful leadership implies that you succeed in creating an environment where people flourish and can perform at their best in a way that they not only realize their own goals but at the same time also contribute to your goals or those of the organization that you are leading. Obviously, leadership belongs on the "people–feeling" side of the scheme, although it relates to every aspect of scientific life. Leadership rests strongly on effective communication and on organization as well. We have included both in Fig. 2.7.

Now let us try to come to a description of leadership. Formal definitions have been given by Locke (1991): "leadership is the process of inducing others to take action toward a common goal"; and by Conger (1992): "leaders are individuals who establish direction for a working group of individuals, who gain commitment from these group members to this direction, and who then motivate these members to achieve the direction's outcomes."

Fig. 2.7: A summary of what scientific leadership entails.

The impact of effective leadership becomes clear in the following statement adapted from Harter et al. (Harter et al. 2002): "Leaders impact the performance of their organizations through their immediate subordinates, whose influence spreads throughout the firm. The personality of the leader affects employee satisfaction, which in turn affects the performance of the organization." It is through motivating the people that the organization becomes successful.

In his 1991 book *The Essence of Leadership*, from which we took the definition above, Locke presents a leadership model based on four ingredients:

- **Motives** (drive and ambition to achieve and willingness to lead and use power to achieve goals) and **traits** (honesty, integrity, confidence, emotional stability, creativity, and flexibility)
- **Knowledge** (content, preferably from experience), **skills** (communication, handling conflict, and building relationships), and **abilities** (managing, planning, decision making, problem solving)
- **Vision** (development and seeking of commitment for it)
- **Implementation** (strategy, selection, and motivation of staff, team building, management)

In this view, leaders distinguish themselves from managers in the sense that leaders establish the vision, while managers implement it. Both qualities are needed; the leadership role can very well include management; the reverse is not true, however.

Having a vision and the inner drive to realize it and the ability to motivate others to commit themselves to implementing the goals that derive from the vision is the hallmark of leadership.

> *"Good leaders create a vision, articulate the vision, passionately own the vision, and relentlessly drive it to completion."*
> Jack Welch, CEO General Electric

### 2.6.1 A working definition for academic leadership

In the context of academia, where education, discovery, and development bring in elements of idealism, of visionary views of the future, and of educating the next generations of leaders, we believe that the following description catches the essence of leadership:

1) Leaders have a vision, a dream of what they want to achieve. They are passionate about it. Their vision places the goals of the organization in a larger context of, for example, the needs of the university, the scientific discipline, industry, or even the society. This vision is shared by all the team members and serves as a source of inspiration and a common basis for a focused effort, without limiting creativity and new ideas. Leaders are experts in their profession and keep a focus on long-term goals.
2) Leaders are great communicators and dispose over emotional intelligence. While cognitive intelligence refers to the objective logic involved in leadership, emotional intelligence is the ability to control one's own emotions and recognize those of others. Emotionally intelligent leaders succeed in creating a motivating/stimulating environment and a social climate in which team members feel at ease to exploit their own strengths. Ideally, there are no barriers for critical discussion and constructive exchange of ideas. Cultural diversity is acknowledged and valued, and the leader and the more experienced group members function as mentors for the younger people. We discuss the concept of emotional intelligence further in Chapter 6.
3) Leaders know the essential management tasks needed to keep the organization running, such as keeping track of the progress of projects and the financial situation, maintaining a proper safety regime, ensuring that reports are ready in time, etc. Of course, they may have delegated many organizational and management duties to others, but they are periodically updated to be well informed about the actual position of the organization. We deal with some of the most important skills for academics in Chapters 3 and 4.

### 2.6.2 Some leadership styles

Thus far, we have defined the ideal leader as an all round manager who masters his/her discipline, has an inspiring vision of where to go, and possesses emotional intelligence. Does that apply to each and every leader? No, of course not. Rooke

Tab. 2.2: Leaders and their main characteristics.

| Type | Characteristics | Strengths | Weaknesses |
|---|---|---|---|
| Alchemist (1%) | Generates social transformations. Reinvents organizations in historically significant ways | Leads society-wide change | None |
| Strategist (4%) | Generates organizational and personal change. Highly collaborative; weaves vision with pragmatic, timely initiatives; challenges existing assumptions | Generates transformation over short and long term | None |
| Individualist (10%) | Operates in unconventional ways. Ignores rules he/she regards as irrelevant | Effective in venture and consulting roles | Irritates by ignoring key organizational processes and people |
| Achiever (30%) | Meets strategic goals; promotes team work; juggles managerial duties and responds to outside to achieve goals | Well suited to managerial work | Inhibits thinking outside the box |
| Expert (38%) | Rules by logic and expertise. Uses hard data to gain consensus and buy-in | Good individual contributor | Lacks emotional intelligence and respect for the less experienced |
| Diplomat (12%) | Avoids conflict; wants to belong; obeys group norms; doesn't rock the boat | Supportive glue on teams | Can't provide painful feedback or make hard decisions |
| Opportunist (5%) | Wins any way possible; self-oriented; manipulative; "might makes right" | Emergencies and pursuing sales | Few people want to follow them long-term |

and Torbert (2005) presented an interesting classification of several types of leaders and their action logics, with their strengths and weaknesses, which we summarize in Tab. 2.2.

**Leadership in Action: The Classical Symphony Orchestra under the Maestro**

*Music ensembles present instructive examples of leadership situations. Let us take a symphony orchestra as an example. Our conductor, the maestro, is an accomplished musician in his late fifties. He has a* ***vision*** *that music is a universal language that unifies people across borders and races; he is passionate about the late romantic repertoire, and as a previous violin soloist, he knows many scores inside out. He prefers to work with young musicians, who he sees as the future generation of conductors to spread his philosophy over the world. The maestro's* ***mission*** *for the coming three years is to make a world tour with his young orchestra and to play in less privileged regions of the world to enthuse a large public for classical music.*

*In the intense rehearsals and concerts, the orchestra feels that the maestro stands above every symphony or solo concerto on the program, and the musicians are eager to play under his guidance and learn from him. The maestro leaves freedom to soloists in the interpretation of solos. During concerts, the orchestra is full of concentration and becomes ever more confident. Occasionally, something goes wrong, but then the conductor is the one who, with calm but clear gestures, gets everyone back on track. The joy of making music is visible and radiates over to the audience.*

*This orchestra has almost become a self-organizing entity, and the maestro can fully concentrate on giving the music just that extra touch that makes it really special. Performances are intense, sometimes exhausting, events where the audience generally feels enchanted. Musicians who played under the maestro are generally proud to mention this in their CVs.*

*The maestro and his orchestra demonstrate leadership in action, in combination with self-leadership of the orchestra, where different sections, e.g., strings, brass, wind, or percussion sections, function as optimal teams. The entire operation runs smoothly, although all concerts are different.*

## 2.7 Concluding remarks on leadership

In summary, successful leaders have a compelling vision that they share with their team, implying that they are experts in their subjects, posses significant emotional intelligence, and are good in establishing and maintaining relations, and they have the necessary management skills, although they may have delegated many organizational duties to team members. We end this chapter with a paraphrased quote from Kathleen Schulz, Director of the American Chemical Society, who in a column of *Chemical & Engineering News*, gave a concise description of leadership (Schulz 2016):

> *"Great scientific leaders create a climate in which their team can do great work. Great leaders are passionate, take time to reflect, are always optimistic, are great communicators, and keep focus on long-term goals."*

We could not agree more!

The archetype of the alchemist and charismatic leader is Nelson Mandela (1918–2013), a leader of society-wide change in South Africa and admired and beloved all over the world, unparalleled in how he stood above the parties and in the end even respected and loved by his former enemies. We list some inspirational quotes:

***"May your choices reflect your hopes, not your fears."***

***Do not judge me by my successes, judge me by how many times I fell down and got back up again.***

***"Its better to lead from behind and to put others in front, especially when you celebrate victory when nice things occur. You take the front when there is danger. Then people appreciate your leadership."***

***Nelson Mandela***

## References

Blanchard, K.H., Hawkins, L., Fowler, S.: Self Leadership and the One Minute Manager: Increasing Effectiveness through Situational Self Leadership; HarperCollins Publishers Inc., New York, 2005.

Conger, J.A.: Learning to Lead: The Art of Transforming Managers into Leaders; Jossey Bass, San Francisco, 1992.

Drucker, P.F.: Managing Oneself; Harvard Business Press, Boston, 2008.

Hay, L.L.: You Can Heal Your Life; HayHouse, New York, 1984.

Harter, J.K., Schmidt, F.L., Hayes, T.L.: Business-unit relationship between employee satisfaction, employee engagement and business outcomes: a meta-analysis; J Appl Psychol; 2002; 87; 268–279.

Locke, E.A.; The Essence of Leadership; Lexington Books, New York, 1991.

Rooke, D., Torbert, W.R.: Seven Transformations of Leadership, Harv Bus Rev 2005; 83(3), 66–76.

Schulz, K.: Chemical & Engineering News, January 25, 2016.

# 3 Presenting science: talks, publications, posters, and some ideas on conferences

## 3.1 Introduction: how to communicate effectively

***"Communication is an act of will directed toward a living entity that reacts."*** This crisp and clear definition of communication comes from Garcia, who bases his theory on the principles of war fighting. Let's analyze his definition further by using a paraphrased quote from Garcia's book, *The Power of Communication* (Garcia 2012):

> ***Communication is an act of will...***
> *Effective communication is intentional. It is goal-oriented. It is strategic. Unlike ineffective communication, effective communication is not impulsive or top-of-mind. Communication is not only about what one says, but about anything one does or is observed doing. It is about any engagement with an audience, be it a group of people or a single person, and it includes silence, inaction, and action.*
>
> ***...directed toward a living entity...***
> *Listeners are not passive bodies that absorb messages. Rather, they are living, breathing human beings. They have their own opinions, ideas, hopes, dreams, fears, prejudices, attention spans, and appetites for listening. Most important, it is a mistake to assume that audiences think and behave just as we do. Most do not. Understanding an audience and its preconceptions, and the barriers that might prevent an audience from accepting what one is saying, is a key part of effective communication.*
>
> ***...that reacts.***
> *This is the element most lost on many leaders. The only reason to engage an audience is to change something, to provoke a reaction. Ineffective communication is not noticed, or it confuses, or it causes a different reaction than the one desired.*

## 3.2 A crystal-clear message for a specific audience

Scientists need to be excellent communicators to be successful. They present their work, their ideas, their plans, and their courses

- in spoken word at conferences during lectures or posters sessions, in seminars, group meetings, student information events, in pitches during match making events, in the media, or in interviews with a journalist; and
- in written text in articles, reports, lecture notes, posters, and not to forget grant applications, necessary to obtain the funding for their activities.

https://doi.org/10.1515/9783110468892-003

When using PowerPoint slides or when explaining a poster, written and spoken words together form the most obvious content of what is presented. But there are many more elements in a presentation that matter: think about intonation, facial expression, body language, the way you respond to the audience, in answering questions or just in how you behave when a mobile phone goes off.

In written text, e.g., articles, reports, applications, lecture notes, brochures, and web pages, each medium requires its own approach in terms of style, degree of formality, level of detail, balance between text and illustration, etc.

All this constitutes your side of the presentation, and you will be most successful if your presentation, report, or application is centered around a crystal-clear message that cannot be misunderstood.

The second factor is the receiving end of the communication, i.e., the audience or your readers: living entities that not simply absorb everything they hear or read, according to Garcia. Who are these people who are listening to you or are working themselves through your texts? What is their background? How do they take up information? How much time and effort are they willing to invest to get your point? How easily is their attention distracted? What causes this distraction? Is it you or your style of presenting or writing? Are there external causes, like construction activities in the next classroom, telephones ringing in the audience, two people whispering in the last row of the lecture hall, a deficient sound system, or faltering electronics? Or in written text, is it the way you write, the layout of the text, the use of (or lack of) figures, tables, etc.?

Analyzing your audience or readership is very important, so that you can target your message at the right level. At the same time you should do everything to make your message as clear as possible, and to present it as clearly as possible.

Our philosophy about presenting science is very simple:
1) Feel free to develop a style that fits you best.
2) Avoid the obvious mistakes that speakers and authors so frequently make.
3) Always build your speech, report, article, or poster around a crystal-clear message tailored to the people you want to reach.

## 3.3 Writing publications

*Manuscripts that address a relevant question and present a clear message based on original and new results, written in the correct format and in clear English, will always be published, but not always by your favorite journal.*

### 3.3.1 Why do scientists publish?

That sounds like a silly question, isn't it? Maybe you are a young PhD student who just reached an exciting conclusion of your first successful research project. Naturally, you can't wait to see your work published, preferably in a highly ranked journal that is widely read by your scientific community.

**Why scientists publish**
**Scientific reasons**
- To present new results, methods, insights
- To summarize the state of a field (perhaps from a specific perspective)
- To "claim" a subject (establish precedence)

**Strategic reasons**
- To maintain funding (external and internal)
- To obtain status (personal, institutional)
- To qualify for a PhD degree, promotion, tenure.

Through publishing, you share your important findings with scientists all over the world. Simultaneously, you hope that you will become known for this success and that it may help you to build a successful career. Maybe your university requires publications before you can submit your PhD thesis. Your supervisor will obviously be happy with the new insights gained, but he/she will also be happy that the research group can boast another good article on the publication list. It will be difficult without scientific production to acquire funding for the research. The university wants to see publications to prove its status as an important research institute. And so on. Hence, a successful publication is important in many ways, not only for its content per se.

A study by one of the largest scientific publishers, Elsevier, conducted in 2005 revealed that authors still find the dissemination of new results the most important reason to publish, but factors such as furthering one's career, maintaining funding, gaining recognition, or establishing precedence as expert on a topic were very important as well as a second motivation. The consequence is that authors and their institutions are not just satisfied to see their work in print, they also want to see it in a journal that has prestige in the specific field of the author or rather even in the entire scientific community.

### 3.3.2 Measures for prestige: Hirsch-index and impact factor

The scientific community has relatively recently (i.e., over the past 10–15 years) adopted two "quick and dirty" indicators for standing in the field, namely, the Hirsch-index (H-index) and the impact factor (IF). Neither of the two is normalized, e.g., for age and experience of the author or for size of the scientific domain or readership, implying that great care should be taken to interpret what these indicators mean. Both are based on citations, i.e., references to articles by other published articles in recognized journals (books, reports, patents, abstracts for conferences, lectures, websites, etc., do not count). It is good to realize this limitation: parameters for the impact of scientific work are not based on real usage but solely on citations in scientific journals, which is not necessarily the same!

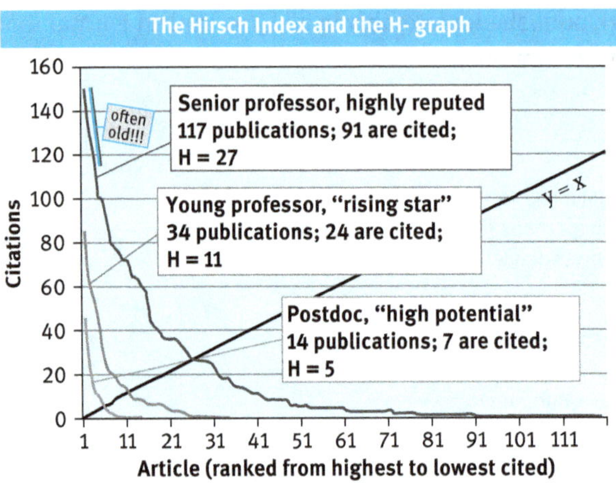


Fig. 3.1: The H-index and some examples.

#### 3.3.2.1 The H-index

The H-index is the number, H, of articles of an author (primary author or coauthor; there is no distinction in the roles that coauthors have played) that received at least H citations.

The way to determine someone's H-index is to rank his/her publications in decreasing the order of citations and see at which number in the list the number of citations is less, as graphically illustrated in Fig. 3.1.

**Publication lists do not tell it all**

When starting his PhD program, Jean-Pierre joined an ongoing and successful research program under a famous supervisor. He could almost immediately start to carry out the ideas suggested by his supervisor, and experienced students showed him exactly how to use the equipment. He gathered data for five good level publications, together with his professor, who was happy with Jean-Pierre's performance. The papers were cited from the beginning, and when Jean-Pierre left, his H-factor was already 5 and on the rise. However, how much did Jean-Pierre really learn? Indeed, his next job, research scientist in an institute, was not a particular success, as he had to build up his own laboratory. As interacting with other people was not one of his strengths, he did not get much done. His director terminated his contract after 1.5 years.

Javier was the first PhD student in the group in a new research program. He had to order and build up equipment and test it and was very happy that in the final year, he could do some successful experiments and write two publications. It took some time before these novel studies started to be noted and cited. His H-index is clearly lower than that of Jean Pierre. Nevertheless, Javier is much better suited for a research career than Jean-Pierre is, who still has to demonstrate that he can build up an activity by himself. Good recruiters will recognize the difference, but unfortunately, in the present competitive climate, Javier may have more difficulty finding an academic position than Jean-Pierre, should they want to go in that direction.

To illustrate why the H-index is a "quick and dirty" parameter, one simply should note that an H-index of 10 implies that the author will have at least 10 papers and 100 citations in total, as a minimum. Suppose he/she has colleague who also published 10 papers, with 5 of them hugely successful (50–100 citations per paper) and the other ones below 5; his/her H-index is still 5. Nevertheless, the latter colleague has by far the highest impact with his/her papers, though his/her H-index is lower. Another peculiar point is that any scientist who stops publishing after a few years will see his H-factor go up as long the work remains cited. Some of our colleagues jokingly say that one should strive for an H-index equal to one's age. That could prove quite a challenge! Maybe half one's age is a nice target, or years of research activity? Don't take the H-index too seriously, just publish high-quality papers and explain to the bureaucrats of your university that these simplistic indicators hide a lot of important nuance.

#### 3.3.2.2 The IF (and the citation half-life)

Prestige is based on how much your work is used by others, as measured by citations, also for journals. Publishers want a quick indication of success, and prospective authors want to know how well a specific journal is read, in the hope that their work has a better chance to be cited by others. Ironically enough, with all the search engines available these days, it is less important where a paper is published, as readers seldom read through hard copies of journals anymore. Nevertheless, the IF has become an indicator of prestige, perhaps even more so for the administrators who keep track of how the university is doing than for authors themselves.

The IF for a specific journal is the total number of citations to articles in that journal in that year divided by the total number of articles published in the two previous years. Typical numbers are on the order of 1–5 for topical journals (i.e., of a specialized sub-discipline), 4–10 for general journals, 8–15 for reviews, and 20 or higher for prestigious journals like *Science* and *Nature* and some of their spinoffs.

There are several complications with the IF that users should be aware of. For example, the IF concept does not account for the size of a field and how large the typical readership of the journal is. Another consequence of the definition is that as the scientific community grows and tools for online search and for swiftly including references in manuscripts become more widespread, many IFs tend to rise over the years.

Also, a period of minimal one year and maximal two years is very short for collecting citations, as a publication often has some induction time before it starts to be known and cited. To circumvent this, some journals define another IF on a five-year basis; interestingly, the values are not very much different from the usual two-year values.

Sometimes, publishers try to manipulate IFs by artificially releasing issues in the beginning of the year (pre-publish papers in the Fall with a formal publication in January thereafter) or to systematically publish a review of widespread interest in the first issue of a year. Some journals encourage authors to include at least three references to recent articles in the journal they are submitting to, which is a questionable practice, of course.

In this sense, the citation half-life is probably a more reliable indicator of a journal's quality, as a long half-life indicates that it publishes work of long-lasting value. Hence, typical archival journals can boast citation half-lives of 10 years or more. The same is true for review journals (the well-established ones have both high IFs and long citation half-lives).

Why is a parameter that measures how soon and often the articles of a journal are cited perceived as a mark of quality for authors? Because a high IF is regarded as an indication of prestige, and hence, getting accepted in such a journal is regarded as an accomplishment. We see more and more that authors in their curriculum vitae (CV) include IFs for the journal in which they published. Again, experienced scientists are hardly impressed and focus on content, but university administrators and recruitment officers seem fond of these simple indicators.

In conclusion, prestige indicators such as H-indices and IFs are easy-to-calculate numbers based on number of citations in recognized scientific journals, but one must be aware that they disguise a lot of detail and therefore should be used with great care.

### 3.3.3 The basic principles: a clear message for the readers

Scientific writing follows simple principles: the author has a clear message and he/she wants it to be read. Readers want a crystal-clear and easy-to-digest message about an issue in which they are or may become interested. To this end, articles follow a more or less standardized format, in which readers know where they can find what they want to know.

In an era where publications have almost become "scientific currency" and the fact of having them is almost as important as their content, it is good to realize why we do publish. In his book *How to Write a Paper*, Richard Smith (1994) poses the following questions to prospective authors who are about to write a manuscript:

1) What do I have to say?
2) Is it worth saying?
3) What is the right format for my message?
4) What is the audience for my message?
5) What is the right journal for my message?

Journals and their readership expect a focus on quality and novelty. In Tab. 3.1 we quote from our distinguished colleague Professor Roel Prins (ETH Zurich; long-time editor of the *Journal of Catalysis*), who taught many courses on scientific publishing:

Tab. 3.1: What quality journals are looking for (courtesy Roel Prins, editor).

| Wanted | Not wanted |
|---|---|
| Originality | Duplication ("me too research") |
| Significant advances | Reports of no scientific interest |
| Appropriate methods & conclusions | Work out of date |
| Readability | Inappropriate methods/conclusions |
| Studies that meet ethical standards | Studies with insufficient or incomplete data |

Another relevant quote with respect to strategic publications, e.g., those of authors seeking to fill their publication portfolios rather than making a novel contribution, is the following:

> ***"Just because it has not been done before is no justification for doing it now"***
> (Peter Attiwill, Editor-in-Chief, *Forest Ecology and Management)*

If you are convinced that you have something important to publish, you can go ahead and start preparations. But first it is good to review once again what scientific writing is about (see Tab. 3.2). We will now go through the stages in preparing a manuscript.

Tab. 3.2: The elements of scientific writing: clarity, objectivity, accuracy, and brevity.

| | |
|---|---|
| **Be clear:**<br>– Ordinary, short, familiar, non-technical terms are better than long, grand, unfamiliar technical and abstract vocabulary.<br>– Avoid complicated sentences.<br>– Realize that English is a foreign language for many authors and readers. | **Be objective:**<br>– Be unbiased, unemotional, and truthful.<br>– Give evidence to support your arguments, and acknowledge merit in other views.<br>– Separate results from interpretation.<br>– Benchmark results against the best available information in the literature. |
| **Be accurate:**<br>– Facts need to be accurate, complete, and verified more than once.<br>– Avoid ambiguous or misleading statements.<br>– Carefully check all data and procedures that are reported.<br>– Include indications of accuracy and reliability, such as error margins, standard deviations, etc. | **Be brief:**<br>– Use efficient word usage but avoid uncommon abbreviations and jargon.<br>– Give a concise Introduction.<br>– Give a brief discussion<br>– Provide effective, self-explanatory illustrations, with informative captions, and clear tables.<br>***A high-quality picture is worth a thousand words*** |

Jilin, a PhD student at an Asian university, had little success with her research so far. She could, for her thesis, only write chapters on introducing her subject and on the experimental procedures she had used in her research. Together with her supervisor, she decided to select a publication from the literature and to repeat the work with a somewhat altered composition of the central material – a catalyst – but keeping the essential ingredients the same. The "me-too approach" appeared successful and Jilin obtained slightly different results as in the paper she used as inspiration, although not so that conclusions would have to be changed. Jilin wrote a thorough paper, which she discussed at length with her supervisor, and she confidently submitted it to a well-established journal of the field. However, the editor returned it after a few days. He found the paper technically sound but did not feel that he learned something he did not know yet. Jilin was disappointed but submitted it to another journal, where she initially had more luck. The editor sent it to three reviewers. One recommended publication after some revision, the second had some doubts about novelty and advised to shorten the manuscript substantially, while the third reviewer happened to be the author of the original paper Jilin based her work on. He was not impressed at all and recommended rejection due to lack of originality. The editor had little other choice than to refuse the paper. Finally, Jilin managed to get the paper published in a new open-access journal that was not so critical in its selection procedures yet. So, in the end, Jilin had her published article for inclusion in her PhD thesis, but not at the level she had hoped for. Of course, we sympathize with Jilin and are happy that she succeeded to graduate. However, we strongly disagree with university policies to require published articles before a student can graduate. In the case of Jilin, she clearly demonstrated that she can perform state-of-the-art research at the international level. That it was not at a particularly high degree of novelty is something the supervisor should be concerned about. We find it an inappropriate requirement for a young student, whose main interest is to learn from the experience.

### 3.3.4 Stage 1 – Before you write

#### 3.3.4.1 Study the Guide for Authors

When you know what you want to publish, and where, then our first recommendation is to read the Guide for Authors of the journal you have in mind. You would be surprised to know how many inappropriate manuscripts journals are receiving. These are mostly rejected right away simply because authors failed to check the scope of the journal and the requirements for submitting a manuscript. Also, should the journal of first choice reject your paper, then never submit the same material in the same form to a second journal without checking its Guide for Authors first!

#### 3.3.4.2 The message in one sentence

Try to capture the essence of your paper in a single sentence. This sounds simple, but it is not. In a sense, it is the analogue of presenting your message in 30 seconds, sometimes referred to as the "elevator pitch" (see Chapter 4). However, if you manage to catch the intention and conclusion of your article so concisely in one sentence, it means that you know exactly what the focus of the paper should be and what needs to be included.

An example for such a one-sentence message (basis for a paper in the *Journal of Physical Chemistry*; we apologize for the highly specialized nature of the content), is as follows:

*Polyethylene formation from ethylene with a chromium catalyst occurs on atomically dispersed, divalent chromium ions coordinated to the silicium oxide support via two oxygen ions, as evidenced by data from surface spectroscopy and electron microscopy obtained on a model system for the industrial catalyst.*

Such a "message in one sentence" is your guide to select (i) topics for the introduction, (ii) results, (iii) explanations, (iv) description of methods, and (v) pertinent literature that you need to include for a truly convincing story.

#### 3.3.4.3 Cite the pertinent literature

**Literature from before 2000 is not necessarily outdated**! Young academics sometimes seem to think that science – or at least their discipline – started in 2000 and everything before is textbook material. However, a lot of excellent research and thorough thinking were done throughout the previous century. Methods of investigation and computer power to interpret and simulate measurements may be much more refined nowadays, but many scientists in the previous centuries generally had a very thorough education in the basic disciplines and sometimes very clever ideas. Also, they were under much less pressure to acquire funding and turn out papers. Hence, the articles they did publish were often meticulously prepared and very well thought through. In chemistry and physics, many articles and books from the early 1900s and onward offer exciting ideas and can be a great source of inspiration. Do yourself a favor and take a few hours now and then to browse through old monographs and journals of your field (hopefully, your library still has them, but you can find some famous older literature on line as well).

Next point on the agenda is to review again all related literature and to identify the key papers that must be included in your manuscript. Be careful here; it is tempting to collect a long list of references from a service such as Scopus or the Web of Science and to include all these in your manuscript. However, try to be selective and mention only key papers that you have read or still must read before you finish your manuscript. Keep in mind that it is an ethical obligation of every author to give proper credit to publications you have used in any way, particularly in the design of experiments, in the interpretation of data, and in the construction of theories in the discussion. It is also a moral obligation to check the literature carefully for papers on your subject, and you may have to go back into the previous century (see Textbox). On the other hand, it is not considered proper to cite for purely political or strategic reasons, for example, to help a befriended colleague or to please a person who might be the referee on a future grant proposal. Excessive self-citations fall in the same category; but of course, it is acceptable, and in fact expected, that you refer to your previous work on the subject if it is related to your present manuscript. References should be given in the proper format as requested by the journal, taking care that all names are spelled correctly.

**Strategic citations – not always the author is to blame.**
A questionable example of strategic citations is the following. A recent manuscript from our own laboratory was reviewed for a well-established journal and found worthy of publication after some revision, provided – and this was a request from the editor! – we included more citations to recent work published in his journal. Apparently, the editor found this an effective way to boost the impact of his journal. Such a request is, of course, unethical.

#### 3.3.4.4 Who qualifies for authorship?

This can be a tough question. You – and if you are a PhD student or postdoc, together with your supervisor – will have to decide who the coauthors will be on the paper. Modern science is teamwork, and several people may have given essential input for the research that is the basis of your manuscript. What about indirect but vital contributions, such as the laboratory infrastructure in which you have worked and the funding for the work; should these be considered as justification for participation in authorship? Different organizations have different rules, and your supervisor will probably have a policy for handling such issues.

Our recommendation is to consider only those people as coauthor who made a substantial contribution to the actual work and who are capable to defend at least a part of the paper's conclusions or claims. Furthermore, they should have an active role in either the original writing or the revision of drafts of the article and they explicitly agree to coauthorship of the final version as it is submitted.

Many journals nowadays will ask clarification of the roles of each individual author when you submit the manuscript, but be aware that policies for accepting coauthorship are not uniform.

The order in which the authors appear also requires serious consideration. There are two favored positions: first and last author. Some organizations even value the status of corresponding author; several journals indicate this by an asterisk or footnote. Who deserves the preferred positions? Usually, the person who did most of the work and wrote most of the paper is the first author, while the most responsible author – e.g., often the supervisor – takes the last position. However, some laboratories practice different rules, for example, to always place the most senior professor's name first or last or to put names in alphabetical order. In our laboratory, we have mostly practiced the rule that the student or scientist who does most and writes the manuscript goes first, while the senior scientist who takes final responsibility goes last. Hence, in our publications, you will find the director's name often in the middle, as the younger staff is usually responsible for the project and for daily supervision of the PhD student involved and takes final responsibility for the content of the paper (and automatically serves as corresponding author, whether the role is prestigious or not – for us, it is merely an administrative one).

#### 3.3.4.5 Which journal to choose?

Many authors nowadays use the IF as a guide for selecting journals, or they select one of the well-established journals of their field. Some authors will try to publish as high as possible in a more general journal of high reputation. If rejected, they go to the next one in the – sometimes only perceived – reputation list, until the paper gets accepted. Be aware, however, that most journals have a clearly defined scope, as described in the Guide for Authors, and that the first thing an editor will do is to judge whether your manuscript fits in. You may use the list of references of your manuscript as a criterion to judge if your choice of journal makes sense. If your target journal, or related journals, do not appear in your list, you may have made the wrong choice. Editors often also use this criterion as a quick indicator for appropriateness.

We repeat: whatever journal you choose to submit your work to, make sure that you always read the Guide for Authors. Your manuscript must satisfy the requirements for, for example, layout, section lengths, nomenclature, number and type of figures and tables, and reference style. Do not annoy the editor by neglecting either the scope of the journal or its requirements for layout.

#### 3.3.4.6 Ethical obligations of authors

Unfortunately, the world of scientific publishing has seen quite a few cases of misconduct recently, ranging from questionable practices such as duplicate publication, self-plagiarism, or violations of good practice by not revealing complete sample information on materials or methods, to outright fraud, such as falsification and plagiarism. With respect to scientific articles, the following list (Dodd 1997) summarizes the main ethical principles for authors:

- Present an accurate account of your research and an objective discussion.
- Provide sufficient detail and references to enable repetition of the research.
- Identify hazards.
- Cite the pertinent literature and avoid excessive self-citation; identify the source of all information quoted, except what would generally be considered as common knowledge.
- No unnecessary fragmentation of research reports.
- Inform the editor of related publications under consideration elsewhere.
- No duplication of previously reported work (except in a review).
- Criticism of other work is permitted, but never criticize persons.
- Coauthors share responsibility and accountability for the contents.
- Mention sponsoring parties and disclose any possible conflict of interest.

True scientific leaders will not only adhere strictly to ethical principles themselves but also teach these actively to their students and coworkers.

### 3.3.5 Stage 2 – Writing the paper

#### 3.3.5.1 Work from a basic structure; preferably separate Results and Discussion

After you are clear about the message of your article, and you know what its format will be, comes the moment to sit down and start writing. It is good to have a basic idea of what the eventual length is going to be. Table 3.3 lists the well-known basic report structure that most journals adhere to for regular articles (letters and short communications usually have less or no sections). We have also included suggestions for the length of the manuscript as a start. Of course, you use as many pages as it takes, but try to keep the manuscript concise, and do not write overly long introductions and discussions. Also keep the number of references to what is needed, and don't inflate the list (see above: cite the pertinent literature).

Nowadays, the trend is toward shorter papers with more material in a Supplementary Section. However, the risk is that many readers will never look at this section, which implies that material you place here has a big chance to go lost. We therefore recommend to keep this section to a minimum and avoid using it to publish important results that would be valuable for their own sake.

Unfortunately, another trend is to more and more mix results with discussion. Our strong recommendation is to write separate Results and Discussion sections. The main value of your paper is in the novel data that you are reporting. If documented properly, these results are of long lasting value. However, you will be interpreting their meaning in a context of what is known *now*. Insights usually change over time. If your results are presented within an obsolete framework of thinking, they will probably not be used anymore. Results, provided they are presented well and properly documented with the conditions under which they have been obtained, can have eternal value; interpretations may be outdated tomorrow.

In the same vein, accurate and complete descriptions of the materials used and the experimental procedures employed are essential for the value of your article. It is an ethical obligation of authors to present the work such that peers in the same discipline can repeat the work. Hence, leaving out crucial details is not allowed, although it happens unfortunately.

Tab. 3.3: The structure of articles and reports.

| **Sections and suggested length (pages)** | |
|---|---|
| Abstract | 0.5 |
| Introduction | 1–2 |
| Experimental methods and materials | 1–2 |
| Results | 3–6 |
| Discussion | 2–4 |
| Conclusion | 0.5–1 |
| Acknowledgments | <0.5 |
| References | As needed |
| Supplementary material | As needed |

#### 3.3.5.2 What to do about writer's block

You started writing, and you finished the first lines, which took longer than expected. Even worse, the next line even takes more effort to get it on the screen. Although you formulated your "message in a sentence" (see above) and started with a clear idea about what needs to be written up, the right words don't seem to come. Every sentence you write looks bad, inaccurate, poorly formulated, and does not quite express what you wanted to say. This happens often.

Chances are big that you are suffering from writer's block. It is a common phenomenon, particularly among less experienced authors. The basic reason behind writer's block is that you are trying to achieve two things at the same time, which by nature don't go along too well:

- Getting your ideas on paper is an act of creativity; it uses the right side of the brain, where intuition, creativity, and appreciation for art and music reside.
- Getting sentences in proper form, however, is a skill-based effort; it uses the left side of the brain, where analytic capabilities, logic, language, scientific, and mathematical skills reside.

Combining creative ideas with technical skills to express an idea in a perfect sentence implies that you block one activity by the other. Hence, learn to apply the following principle:

***Write first – Then get it right***

In other words, write down what comes to mind, and don't worry about language, correct grammar, sentence structure, or spelling. Tomorrow, you can revise your text and get it in proper shape.

Another advice that will help you is to feel free to start in any section where you have the feeling that you know what to write, or even to add in random order to different sections. In other words, write in a convenient order. Here the recommendation is also to write first, and then get it right. See Tab. 3.4 for a suggestion, but do it as it works best for you.

Tab. 3.4 Writing the manuscript.

| **Write in a convenient order (and feel free to do it your way)** | |
|---|---|
| **Experimental** | Brief but complete |
| **Results** | Clear figures, informative captions, complete tables, then describe them |
| **Discussion** | Summarize key results, asses their value, place in context, discuss significance |
| **Conclusion** | In essence, an elaborate version of your "message in a sentence": Start with the aim of the work, then list the conclusions, followed by one or more sentences of perspective on what next |
| **Introduction** | Context, aim of your work, key literature, identification of a clear research question and your (novel) approach, preview of the message |
| **References** | Use reference software; limit references to ones that are needed |
| **Title** | Short form of your message, specific, concise, inviting to read |
| **Abstract** | What, why, how and significance, clear message |

#### 3.3.5.3 Attractive illustrations and tables with understandable captions tell a story

Figures are extremely important and should be laid out with care. Often, however, the figures we see in publications are far from ideal. Please realize that the common software for presentations and spreadsheets *don't at all* produce good figures. Place carefully designed texts on ordinates, insert labels on spectra and curves inside the figure and not in legends or captions, and use labels that are understandable and not some type of secret code that only the author understands see Fig. 3.2. Figure captions should be clear, informative, and attractive, such that the figure can be understood without reference to the text.

The same principles hold for tables; also, these should be able to stand on their own and tell a part of the story. Every author can make high-quality illustrations, but it takes effort (which is well invested!). Just realize that several readers will not have time to spell out the text, but rely on title, abstract, figures + captions, and the conclusion before they may decide to read the entire paper. And initially, even more important, the editor to which you submit may do the same.

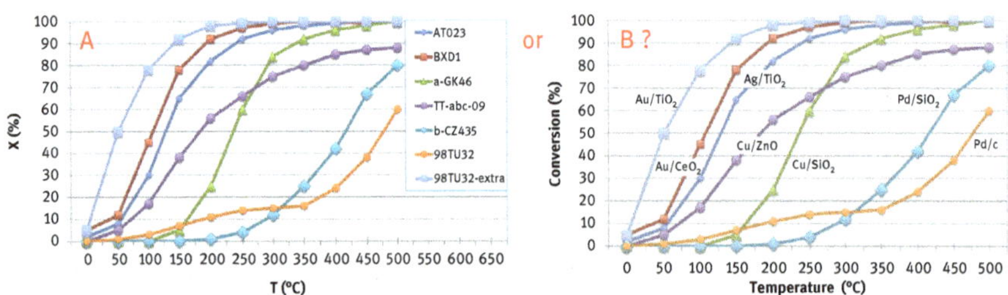


Fig. 3.2: Two examples of how data can be presented: (A) as taken directly from a spreadsheet with measured data, along with experiment codes from the laboratory and (B) the same graph but with understandable labels on axes and data. If you present graph A to a reader, he/she will mostly be busy with looking back and forth between the curves and the codes in the legend while trying to remember what these secret codes stand for. If you show the right version, the reader can immediately concentrate on the meaning of the data, as the figure largely explains itself (the graph shows the degree of conversion of a certain chemical reaction in combination with different metal catalysts on support materials).

#### 3.3.5.4 Sections that often get too little attention: title, abstract, and key words

It sounds strange, but many authors do not spend enough effort on these elements that will be freely available to everyone. Nevertheless, title, abstract, and key words are the first elements on which editors decide if they will consider the manuscript, while later prospective readers use it to decide if they want to see the entire paper.

The title is the main attention getter; it should be brief, specific, and attractive and contain signal words that trigger interest. Constructions such as “A study into the effect of...” essentially waste space and look boring. Think back to the “message in a sentence” – maybe a shortened version could serve as the title? Instead of listing examples, we strongly advise you to browse through some recent issues of your favorite journals and try to identify some effective titles.

Key words are important for abstracting services and for internet search machines and can be essential for prospective readers to find your paper; hence, providing a careful selection of such terms is recommended. Often, "key phrases," which combine a few words, will reflect content more accurately, and many journals allow these.

The abstract is of decisive importance for getting your paper published and read. Editors use it to judge whether your paper fits in the scope of the journal and readers to decide if they will read your paper. Use the abstract to convey your message, and don't use it as a verbose list of contents. An efficient abstract is written around the message of your paper; it explains the significance of the work, motivates why the research has been done, and conveys the major conclusions. Good abstracts are brief, specific, and accurate; avoid hype; use no abbreviations or technical jargon; and cite no references.

Title, key words, and abstract, together with author names and affiliations, form content that appears freely online – visible for everyone, and not only for those with access to the journal. Hence, give these crucial elements the attention they deserve.

### 3.3.6 Stage 3 – Submission of your manuscript

Nowadays, practically all manuscripts are submitted online and no longer by mail. Many publishers have perfected their submission portals to levels that the actual process of submission can be a breeze, provided you are well prepared and have all requested documents ready. Carefully check the Guide for Authors on what is needed. Incomplete submissions are generally not considered and returned to the author.

#### 3.3.6.1 Convince the editor to consider your paper: the submission letter

Write a clear and concise submission letter to the editor. Your aim is to convince the editor that your paper is worthy of consideration for the journal and that it should be sent out for peer review. Check if the Guide for Authors gives any requirements on submission letters and then carefully draft it, addressing in any case the following points:

- The message of your paper and why this message is important, novel, and deserves publication.
- A motivation for choosing this journal; trying to flatter the editor by praising the high IF will not convince him – referring to the scope of the journal and a history of other papers on related subjects that the journal published will be more appropriate.
- If you have coauthors, provide a statement that all agree with the content and have given their permission to submit the work. Often, a short explanation of each author's role is requested by the journal.

- A statement that the work has not been published, even not in part, and is presently not considered by any other journal. If you have submitted a related manuscript elsewhere, you must mention this and include a copy for the editor's information, so that he/she can verify that there is no overlap or that the main conclusions have not yet been given away.
- A few suggestions for reviewers who can give an objective expert opinion on your manuscript. Avoid mentioning colleagues from your own institution or coauthors from your recent papers, and it is also not helpful to suggest the leading star scientists in the field. The editor may invite one of the people you suggest, but more likely will he/she add these names to his database for future use. It is acceptable to also mention names of individuals who should *not* be invited for review due to conflicts of interest.

### 3.3.7 Stage 4 – The editorial process: editors and reviewers

The editor is the intermediate between author, publisher, and the scientific community, a sort of "broker" who understands the interests of all parties involved. Associate or assistant editors can assist the editor in chief. Editors are almost always experts with considerable seniority and recognition in the international community of their field. They have extensively published themselves and reviewed many articles. Their stature in the scientific discipline gives them the authority needed to take important decisions

- on the scope of the journal (usually together with the editorial board of equally well qualified experts from the field);
- on the manuscripts that have been submitted to their journals; and
- on matters of ethics and scientific misconduct.

Figure 3.3 gives an impression of the timing in the handling of manuscript, taken from the work flow of a typical, well-established topical scientific journal publishing several hundred papers per year and with an IF in the range of 5–7.

#### 3.3.7.1 Your paper has been received and receives a first decision

Normally, a support office at the publisher's assists the editors. This office commonly serves several journals. When your paper is received, the office will log it in the system, check if the submission is complete, ask you for additional information if needed, and finally will forward it to the editor.

The editor then applies a first screening, implying he/she briefly assesses the subject, the quality, and the novelty of the manuscript and he/she decides whether it fits in the journal's scope and whether it meets the requirements for quality of presentation. If positive, the editor may then decide to delegate the manuscript to an associate editor for further handling or to send it out for review immediately.

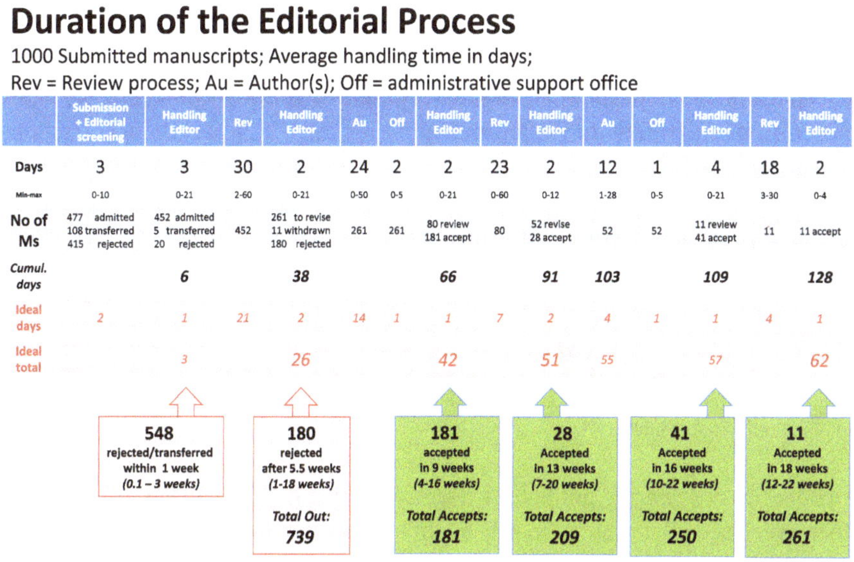

## Duration of the Editorial Process

1000 Submitted manuscripts; Average handling time in days;
Rev = Review process; Au = Author(s); Off = administrative support office

| | Submission + Editorial screening | Handling Editor | Rev | Handling Editor | Au | Off | Handling Editor | Rev | Handling Editor | Au | Off | Handling Editor | Rev | Handling Editor |
|---|---|---|---|---|---|---|---|---|---|---|---|---|---|---|
| Days | 3 | 3 | 30 | 2 | 24 | 2 | 2 | 23 | 2 | 12 | 1 | 4 | 18 | 2 |
| Min-max | 0-10 | 0-21 | 2-60 | 0-21 | 0-50 | 0-5 | 0-21 | 0-60 | 0-12 | 1-28 | 0-5 | 0-21 | 3-30 | 0-4 |
| No of Ms | 477 admitted<br>108 transferred<br>415 rejected | 452 admitted<br>5 transferred<br>20 rejected | 452 | 261 to revise<br>11 withdrawn<br>180 rejected | 261 | 261 | 80 review<br>181 accept | 80 | 52 revise<br>28 accept | 52 | 52 | 11 review<br>41 accept | 11 | 11 accept |
| *Cumul. days* | | *6* | | *38* | | | *66* | | *91* | *103* | | *109* | | *128* |
| Ideal days | *2* | *1* | *21* | *2* | *14* | *1* | *1* | *7* | *2* | *4* | *1* | *1* | *4* | *1* |
| Ideal total | | *3* | | *26* | | | *42* | | *51* | *55* | | *57* | | *62* |

| | | | | | |
|---|---|---|---|---|---|
| **548**<br>rejected/transferred<br>within 1 week<br>*(0.1 – 3 weeks)* | **180**<br>rejected<br>after 5.5 weeks<br>*(1-18 weeks)*<br><br>*Total Out:*<br>*739* | **181**<br>accepted<br>in 9 weeks<br>*(4-16 weeks)*<br><br>*Total Accepts:*<br>*181* | **28**<br>Accepted<br>in 13 weeks<br>*(7-20 weeks)*<br><br>*Total Accepts:*<br>*209* | **41**<br>Accepted<br>in 16 weeks<br>*(10-22 weeks)*<br><br>*Total Accepts:*<br>*250* | **11**<br>Accepted<br>in 18 weeks<br>*(12-22 weeks)*<br><br>*Total Accepts:*<br>*261* |

Fig. 3.3: Example of the work flow for a typical scientific journal (in days from the date of submission).

**Single-blind peer review is the norm**

Different systems of peer review exist in theory:

- Open review: when names of reviewers are revealed to authors (and sometimes readers)
- Single-blind review: reviewer names known only to the editor and not to the authors
- Double-blind review: names of reviewers and authors are unknown to both sides.

The double-blind system is often considered as the most ideal, as any prejudice on the authors or their previous work would be excluded. However, in practice, one can almost always recognize at least the main author, e.g., from citations made in the introduction. Therefore, the single blind system is the common mode of peer review. For open review, the fear exists that many invitations for reports could be turned down because scientists are not keen on being identified as the one who recommended negatively (which could have implications when roles are reversed in the future).

If his/her decision is negative, the editor returns the manuscript to the authors, along with an explanation why he/she finds the work unsuitable for consideration. While you as an author will be disappointed, you should also realize that the editor has a good feeling for which manuscripts have a chance to get positive recommendations in the review process. By his/her off-hand rejection, you have a quick decision and can prepare for submission elsewhere, while the editor also ensures that he/she does not unnecessarily invoke reviewers, whose time is precious as well.

### 3.3.7.2 Peer review

Manuscripts found worthy of consideration are sent out to peer reviewers, usually two to four, who are asked to critically assess the manuscript and to report back in a few weeks.

Reviewers are specialists from the field, often well-established scientists, but sometimes also younger postdocs or even PhD students at the end of their projects, who read the manuscript and then give their opinion on the quality and novelty of the manuscript, its suitability for the journal, and they often indicate where corrections are needed.

Peer review is a system for quality control and improvement, from which the entire scientific community benefits. Note that reviewers do not decide on acceptance or rejection, which is the role of the editor.

**Reviewing is a learning opportunity**
Serving as a reviewer for manuscripts, conference contributions, or grant proposals, is a wonderful opportunity to learn. It gives you insight in how your peers do their research, how they think, and plan their next research.

As the quality of scientific publishing depends critically on the mechanism of peer review, every scientist who is or will be an author has the obligation to do a fair amount of reviewing. Writing a balanced report takes time, but it is also a learning experience, and many scientists regard it as a privilege that they have to study certain manuscripts in depth. We therefore recommend young scientists, i.e., PhD students in their last year and postdocs, to become engaged, for example, by asking their supervisors if they can assist and write a report together. This is an excellent opportunity to learn.

Peer reviewers are bound to ethical obligations; we list the most important hereafter (Dodd 1997). Reviewers should

- accept invitations for manuscripts only inside their own area of expertise;
- judge objectively and respect the intellectual independence of authors;
- explain their judgments, if possible supported by references from the published literature, and refrain from unsupported assertions;
- be alert to correct and complete citation of relevant work;
- treat manuscripts fully confidentially, be sensitive to conflicts of interest, and not use information from unpublished manuscripts in their own work, unless with consent from the authors;
- refuse to assess papers from authors who are too close, such as friends, collaborators, or former supervisors ; and
- return their report within 2–3 weeks at most.

#### 3.3.7.3 Editorial decisions based on peer review

When all reviews are in, the editor takes a decision on basis of the reports, which can be any of the following:

- Acceptance (direct acceptance is rare)
- Acceptance on condition that the authors make minor but mandatory revisions (if they do not accept these, they should provide a convincing reason for this),

- Revision required before a final decision can be taken – implying that the revised paper will be reviewed again by the editor or the original reviewer(s)
- Rejection

If reviewer recommendations conflict each other, the editor may invite additional reviewers, e.g., from the editorial board, or he/she may rely on his/her own judgment.

In case a paper is rejected, be it in first or second instance, authors always have the right of rebuttal. A request hereto can be directed either to the handling editor or to the editor in chief. Usually, editors will respond positively to an appeal from the authors, and they will either provide further clarification of their decision or they may decide to request another expert opinion, often from an editorial board member. However, a second decision is final.

Rejection is of course a disappointing outcome, but it happens often, even to well-established researchers and even to world-famous Nobel Prize winners. The important point is to not take such decisions personally but to evaluate honestly why the paper was rejected. Are your conclusions truly warranted, would inclusion of other or more results enhance the chances of acceptance? You always have the right of rebuttal, and sometimes, one must fight for a paper, but often, choosing another journal is the best option.

#### 3.3.7.4 Dealing with revisions

In case the decision is that your paper might qualify after appropriate revision, the author should consider all comments carefully and revise the paper where it is necessary. He/she should also write a reply to the reviewers and indicate the actions that have been taken or explain convincingly why the manuscript was not changed. The experience of many editors is that authors can be quite slow in submitting revised manuscripts; this is often a step causing unnecessary delays in the publication process. A period of 2–3 weeks is acceptable, but generally, it is best to deal with revision as soon as possible.

After submission of the revised paper, the editor may take a decision by himself/herself or send the paper out for review again. Let's hope he/she decides to accept your work now. If not, learn from the comments and submit an improved version elsewhere. If the work has merit, it will certainly get it published!

### 3.3.8 Stage 5 – Promote your work

Congratulations! Your paper has been published. Next, you will ensure it to be read, used...and cited. You should be aware that the average number of citations is on the order of one per paper per year only. Hence, make sure that people will know your work. You may send PDF reprints to selected colleagues in the field, present your work at conferences to promote it, and put an abstract and your best figures on the web, e.g., on social platforms such as ResearchGate or LinkedIn. In any case, promotion of your work is important and citations do not usually come automatically.

### 3.3.9 In conclusion

How do you get your paper accepted? Focus on a clear message based on original results that address a relevant problem or question; stress why you did the work and what comes out; and ensure that you write the paper in the proper format and correct English. Never forget to check the Guide for Authors. Success!

## 3.4 How to give successful presentations

How often have you been listening to oral presentations about interesting science while you nevertheless had difficulty to pay attention until the end? How often did you lose interest or got distracted before the speaker had even come halfway? Was it the subject and content of the talk, or was it the way the speaker presented it?

Many presentations deal with interesting work but are difficult to follow because the speaker unknowingly makes several presentation errors. **By far the largest mistake is that many speakers do not realize how audiences listen.** Go back to the beginning of this chapter and read how Garcia describes an audience as "*living human beings with their own opinions, ideas, hopes, dreams, fears, prejudices, attention spans, and appetites for listening...it is a mistake to assume that audiences think and behave just as we do.*"

**Our advice for giving effective presentations is simple: feel free to develop your own style, but do your best to avoid the obvious mistakes that many speakers make.** This has proven to be a successful recipe. If you really know what errors you should avoid, the chances are high that you will be able to greatly improve the effectiveness of your presentations.

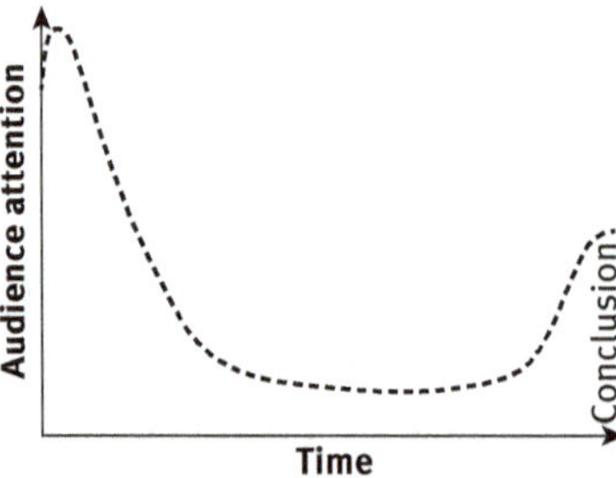


Fig. 3.4: Typical attention the audience pays to an average presentation.

### 3.4.1 The attention curve – help the audience to focus on your message

The average participant in a conference is generally willing to listen to you but also easily distracted. You should realize that only a minor part of the audience has come *specifically* to listen to *your* talk. The rest is there for a variety of reasons, to wait for the next speaker, to get a general impression of the field, or whatever.

Figure 3.4 illustrates how the average audience pays attention during a typical presentation of, let's say, 20–30 minutes. Almost everyone listens in the beginning, but halfway, the attention may well have dropped to around 10–20% of what it was at the start. At the end, many people start to listen again, particularly if you announce your conclusions, because they hope to take away at least something from the presentation.

What can you do to catch the audience's attention for the whole duration of your talk? The attention curve immediately gives a few recipes:

- Almost everyone listens in the beginning. This is *the* moment to make clear that you will present work that the audience cannot afford to miss.
- If you want to get your message across, you should state it loud and clear in the beginning and repeat it at the end.
- The best approach, however, is to divide your presentation in several parts, each ending by an intermediate conclusion (see Fig. 3.5). People who got distracted can always easily catch up with you, particularly if you outline the structure of your talk in the beginning.

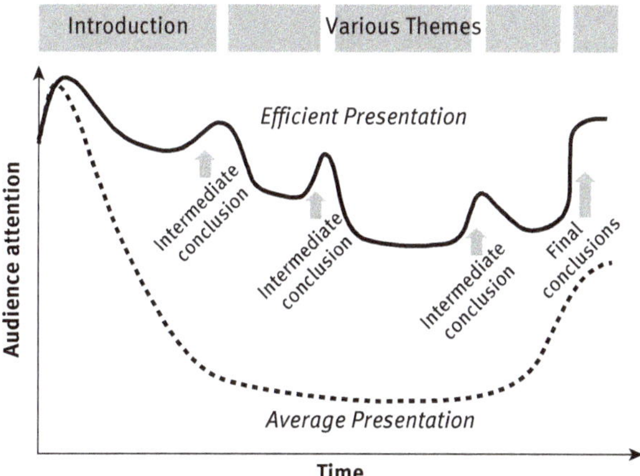


Fig. 3.5: Ideal attention curve of an audience when the speaker divides his/her talk in recognizable parts, each summarized by intermediate conclusions. When people lose their attention for some reason, they can easily catch up with the speaker in one of his/her intermediate summaries. The big advantage of this approach is that every important item is said several times. Repeating the essentials is the key to getting your message across.

### 3.4.2 Why are audiences distracted?

There are many reasons why audiences may be distracted. Some reasons may be outside your control, such as inadequate sound systems, poor projectors, too much stray light from the windows, the catering service entering with fresh coffee and tea, uncomfortable seats with chairs standing too close, or inherently noisy conference centers with sliding walls that partition larger halls so sessions can run in parallel. Bad luck if it happens to you and affects your talk.

What you can do yourself, however, is avoid anything that may encourage the audience to stop listening. Such mistakes fall in two classes: **speaker's errors and presentation errors.**

We list a couple of the most common mistakes, most are self-explanatory.

#### 3.4.2.1 Mistakes you want to avoid

**Audiences love background, but avoid stating the obvious...**
You can raise the interest of attendees who are not especially interested in your subject by giving them the impression that they will learn something from your talk. Note that this part of the audience is more interested in general aspects than in details. You certainly need to give them a good introduction into the background of your subject before they can fully appreciate the subtleties of your work. Hence, you should feel free to spend 20–30% of your time on a proper introduction, without, however, resorting to common knowledge (fossil fuels that will run out, $CO_2$ being a threat to mankind, etc.). You don't want to annoy your audience by stating the obvious. However, with sufficient background information that is to the point, they will understand a lot more about your specific results, i.e., that part of the talk you are most proud of.

- Overestimation of the audience: The speaker lives in his/her own little world of research, believing that all the background information needed to appreciate his/her work is common knowledge. Students often assume that the audience knows as much as their supervisor does, which certainly is not the case! Hence, a topic needs to be properly introduced, depending on the specific audience it is presented for.
- The structure of the presentation is unclear, and consequently, the line of reasoning is hard to follow. Important matters as problem identification, aims, or motivation are insufficiently clear. Sometimes, speakers focus on *what* they have done and *how*, and they forget to tell *Why*. See also the next section on how to organize a presentation.
- Visual aids (PowerPoint slides) are inadequate, too busy, confusing, hard to read, too small, etc. Some speakers show too much in a too short time (one slide per minute is not bad as a rule of thumb).
- The speaker uses long, complicated sentences, or unnecessary jargon, abbreviations, or difficult words. Passive sentences (*"From this figure, it was deduced that ..."* or *"It was therefore concluded that ..."*) sound stiff and are more

difficult to follow than active ones (*"This figure – point at it – implies that..."* or *"Therefore, we conclude that..."*).

- Even worse is when the speaker reads his/her speech from paper and forgets that written language is usually more formal and complex than language used in everyday conversation. Also, reading out written text tends to go a lot faster than telling something in a natural way. Sometimes, politicians suffer from this problem, especially when others have written their speeches. In such cases, the audience will for sure experience information overload, and probably, they will give up and stop listening. Of course, we sympathize with the speaker who feels insufficiently confident in English. However, reading out your text from paper is almost always an unsatisfactory solution. And after all, nobody in the audience will blame you for a couple of mistakes in the language. Don't forget that English is a foreign language for many in the audience.
- Monotonous sentences, spoken either too fast or too slow, lack of emphasis, unclear pronunciation, slang, and colloquial and fashionable expressions (especially if English is the speaker's native language!) all make it difficult for listeners to stay attentive.
- Some speakers turn their back to the audience and watch the projection screen while they are talking, instead of trying to make visual contact with the audience. Sometimes, the reason is that they use the slides to guide themselves through the talk, or they read literally what is written on them. Know that PowerPoint has a presentation mode where one can see the current slide, the timing, plus some speaker's notes on the computer screen while presenting, which is a better solution than looking at the screen all the time. However, never forget to regularly look at the audience and establish contact. Presenting means communicating!

**Not too fast, please....!**

Many speakers have rehearsed their talk so often that in the presentation, they speak too fast. Others simply have so much to cover that the only way to stay within the allotted time is to speed up. Of course, this is not in the interest of the audience, particularly not at an international meeting.

**... and try to vary your pace**

As a rule of thumb, speaking at 150 words per minute is all right. However, try to vary your rate. Key ideas, complicated points, or concluding remarks (you may want to use one at the end of every slide you show) are best presented at a slower pace.

#### 3.4.2.2 How to organize your presentation – not as a written report!

You should be aware of the fundamental differences between an oral presentation and a written report. In a presentation, the listener, by necessity, will follow the order in which the speaker presents the material. The reader of an article or report can skip parts, go back to the materials section, take a preview at the conclusions when reading the results, etc. Exactly because of this reason, all scientific reports follow the generally adopted structure of *Abstract – Introduction – Experimental Methods – Results – Discussion – Conclusions – References*. However, this structure is hopelessly ***unsuitable*** for an oral presentation. Nevertheless, many contributed talks at conferences adhere to it, and several universities actually teach it to their students!

Why is this generally accepted report structure unsuitable for oral presentations? Because the listener will have to remember details about, e.g., the experimental methods until the results are presented, and must recall the various results when the speaker deals with the discussion, etc. In other words, details that should be combined (the why, how, what, and what does it mean, of a specific experiment) are treated separately. You are asking a lot from the audience if they need to remember all these facts and figures, until at the end, you finally explain how these bits and pieces fit in a larger picture!

Grouping together what belongs together is the best principle for organizing your talk. Hence, if you discuss materials characterization, you start this section of the presentation with a few introductory remarks of what you want to learn about your catalyst and how the method of choice may help you to provide this information. Then you show some key results and you discuss their meaning. End with a partial conclusion on this section. Next you go to the following item in your presentation, which may be the chemical reactivity of the materials, or so. When you finished this section, you may give an overall conclusion on the state of your catalyst before you go on to speak about catalytic behavior.

In conclusion, feel free to organize your talk in the way you think it will be understood best, and please, do ***not*** follow the article structure.

### 3.4.3 10 Steps to a successful presentation

The two key issues in preparing a successful talk are the following:
- The message: What do I want the audience to know when I am finished?
- The audience: How do I present my talk such that the audience will understand and remember what I want to tell?

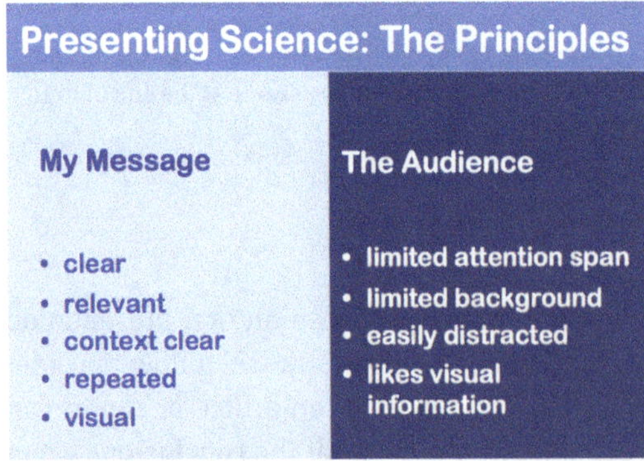


Fig. 3.6: The principles of communication: your message for a specific audience.

**Step 1: Start in time**
Once you submitted the abstract to the conference organizers, it is time to start thinking about how you organize the material in a talk when your abstract will have been accepted. Read about the background of your work, read related work, look at your own results regularly and think about the most relevant conclusions. Try to imagine what type of audience you would have and consider what you would have to include as background information

**Step 2: The message – in – a – sentence**
Try to capture the essential message of your presentation in a single sentence. This is difficult. You will only be able to do this if you really master your subject (which in fact is the main requirement for being able to clearly present your work to others). Such a "message in a sentence" is also a good basis for a 30-second elevator pitch, which we discuss in Chapter 4 (although it is not the same).

**Step 3: Select your material and order it**
Use the "message in a sentence" under step 2 as the criterion to select which results to include, in what order, what basic information is needed to appreciate these results, and which experimental details are necessary and which not. Be very critical, any experiment or result that does not contribute to your main message, however nice it is, should be left out.

Although it may at first sight seem natural to present your results in the chronological order in which you obtained them, this does not have to be the most ideal order for the audience to understand what you have done. Think about where to discuss highlights – at the beginning? Near the end? Maybe you could disperse remarkable features through the entire talk? It is up to you, but apply the order that you believe will appeal most to the audience.

The scientific background of your audience determines how much you should explain about experimental approaches and characterization techniques. Again, be careful *not* to identify your audience with your supervisor; most of your listeners are unlikely to possess much specific knowledge about your subject. By the way, if you explain it well and possibly in an entertaining way, hardly anybody will mind hearing something he/she already knows.

**Step 4: Opening and introduction**
In the opening, i.e., the first few sentences, you catch the attention, for example, by a scientific question or a catchy or maybe even provocative statement. Perhaps, you could already give the conclusion of your work too. Try to speak slowly, with emphasis, and look at the audience. They may need to get used to your voice, your way of speaking. Of course, it is a good idea to prepare the opening carefully, by writing it out in simple sentences, and to rehearse it several times.

However, before you give your opening sentence, it is good to start with "Thank you Mister Chairman, ladies and gentlemen..." followed by a few seconds of silence, in which you look around to establish contact and see if people are paying attention.

By doing so, you implicitly get the audience to listen. Starting this way, you also test the sound system, and you ascertain that your important opening lines are going to be heard. In the rest of the introduction, you sketch the background of your research and you introduce your research question, including the reason why it is important. Remember that many people will be interested in a concise summary of the status in your field. Hence, reserve sufficient time (20–30% of the total time) for the general aspects of your work. It is good practice to not only clearly identify the scientific question you address but also give the conclusion of your work, if you wish so. This way, you enable the audience to better follow your reasoning and to anticipate on the outcome of the experiments. In other words, you give them a chance to listen actively. Remember that a scientific presentation is not a detective story that is solved at the conclusion.

**Step 5: Conclusions and ending**

Conclusions should be properly announced to regain full attention. Present your conclusions in relation to the questions you raised in the Introduction. Avoid all irrelevant details. Once you finished the conclusions, you may acknowledge people who helped you (not the coauthors listed in the program) and the funding agencies. Then comes a final opportunity that many speakers miss: end by repeating the message of your talk, for instance: *"Ladies and gentlemen, I hope I have convinced you that..., etc."* This is the take-home message that the audience should remember, hopefully in combination with your name and affiliation.

**DON'T DO THIS**

An often heard, but poor start is:

*"Good morning, ladies and gentlemen. I am John Talker and* ***today**** *I'd like to tell you something about my PhD project at the Group of Archaic Research at the University of Science City. The title of my talk is.... I will start with an Introduction, then explain the experimental techniques, next present the most important results, and finally I hope to draw a few conclusions and I want to acknowledge a few people. So let us start with the Introduction..."*

If you open this way, you will find yourself in the company of many others. Nevertheless, this is a totally inefficient way to start a presentation. You are simply telling what everybody has read in the program and was already stated by the chairman. How would you respond if you were in the audience?

*Why say "today" in the opening sentence? Do you give another talk tomorrow? Avoid, it sounds silly.

**Step 6: Crystal clear figures and schemes have the highest impact**

A picture is worth a thousand words. Well, not necessarily. Figures, especially those generated by spreadsheets, may look neat and tidy, but at the same time, they may be real puzzles (see Figs. 3.2 and 3.7).

A good picture to be used in an oral presentation

- is easy to read (large lettering, good contrast);
- explains itself (clear title, preferably a conclusion underneath);

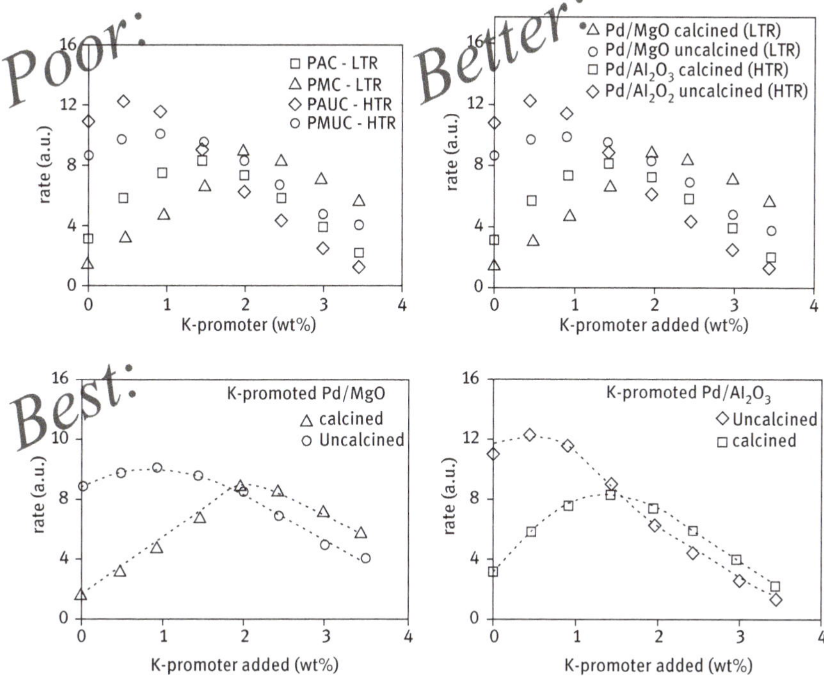


Fig. 3.7: Spreadsheets often produce unsatisfactory figures, particularly with respect to labeling. A good figure has labels on the curves and not in a legend. Secret codes and jargon should be avoided as much as possible.

- contains only relevant information;
- has no jargon or difficult codes that the audience needs to translate; and
- has labels on the data and not in lists in the legend.

Hence, when showing a series of spectra or activity curves, you put an understandable label on each curve (not a, b, and c, which are explained in a separate legend!!). Avoid reference to samples in codes such as "Sample AX234/a5," which may be handy in laboratory notebooks but should be absolutely forbidden in presentations (and in articles as well).

Using tables with numbers is, in most cases, not recommended in a presentation. Remember that an audience will generally read everything you show on a slide, and while they read, they pay less attention to what you say. Also avoid too many theoretical formulas and mathematical derivations. Sometimes, you may have to show one, but try to keep it to a minimum. You should realize that our brains remember graphical and pictorial information best. Hence, clear figures, schemes, and diagrams are the preferred means to convey information.

## Step 7: PowerPoint slides

**Tips for effective slides**

- Use large lettering.
- Aim for contrast: Black letters on a white background, bright yellow on black or dark blue, etc.
- Do not use structured backgrounds and do not waste too much useful space on logos, fancy templates, etc.
- Use pictures, figures, a title, a short, clear caption and conclusion
- Avoid data in tables or inside text
- If you use text, then avoid complete sentences but use "headline" style as the newspapers do.
- Give each slide a meaningful title and try to include a brief conclusion.

PowerPoint is both a blessing and a curse. If used properly, slides are a marvelous aid to present your work. They also leave you the flexibility to make last-minute changes and to adapt to the situation. Unfortunately, many speakers misuse their slides by doing the following:

- Showing too many in a short time
- Overloading slides with data and text
- Using too small fonts
- Showing confusing graphs full of secret codes
- Exaggerating the use of animations
- Forgetting to include a title (aim) on each slide and intermediate conclusions along the way

Also, the usual presentation software offers so many inviting opportunities, that speakers often use ineffective color combinations and disturbing background patterns.

The best advice on making effective slides is probably to remove all information from slides and figures that is not strictly necessary. Nevertheless, do provide clear understandable labels on curves and spectra in figures, so that they become self-explanatory to the audience. Each slide should contribute to the overall message of your talk, and therefore it must be completely understandable.

## Step 8: Think in terms of communication rather than performance on stage

Your presentation will be most effective if you use the same everyday language in which you explain things to a fellow student in the lab. There is absolutely no need to use a more formal language. In fact, formal language is not desirable at all as it is more difficult to understand for the audience. Do not try to impress the audience with fancy words, formal constructions, subject-specific jargon, or unnecessary abbreviations. Think about oral presentations in terms of communication and do not see it as a performance in a theater. The audience will be grateful if they can easily follow what you say, and they will not at all blame you if you make some mistakes in grammar or wording.

**Step 9: Timing is absolutely necessary**

Now comes the moment of truth: does everything you prepared fit within the available time? There is only one method to find out: take your stopwatch and practice it. This is usually a frustrating experience. First, you may note that the sentences simply do not come. As said before, my solution is to sit down and write the first part out in clear, short sentences. Second, you will probably find that you have too much material. Hence, you should cut down, and we do hope that you will not take out too much of the general Introduction. With the attention curves of Figs. 3.1 and 3.2 in mind, it is probably the best to skip a few less important items in the middle of your talk. You should rather not compromise on the Introduction and certainly not on the Conclusions! Carefully timing your presentation is extremely important. Going overtime is an offense to the audience and to the speakers following you, particularly if there are parallel sessions. Nothing is more embarrassing than that the chairman must stop you before you have been able to present your conclusions!

**Don't lose time at the start**

Many speakers, even experienced ones, unnecessarily lose time in the first minutes.

Assuming the chairman introduced you properly, there is no need to restate the title and explain who you are or where you come from. Showing all this information on the first slide is more than sufficient.

Others have noticeable difficulty to get started. Apparently, the intended introductory statements do not come as spontaneously as hoped, maybe because of stage fear.

Note that a good start of the talk is critically important in catching the audience's attention; you don't want to take any risk here. Hence, the best advice to speakers is to meticulously prepare for the first five minutes. Write this part out in short, powerful, crystal-clear sentences and rehearse them several times.

You may do the same for the Conclusions section, these are the last sentences the audience hears and hopefully remembers.

**Step 10: Dealing with stage fear – Are you nervous? Hopefully you are!**

Only very few of us have been born as a talented speaker. Almost everyone will be nervous before a presentation. For beginners, nervousness may easily lead to lack of confidence, caused by feelings of being inexperienced. However, more experienced speakers know that being nervous is a normal part of the game, and nothing to be afraid of.

First-time speakers often interpret nervousness as a sign that they are apparently incapable of delivering a good presentation. This is not true. All the symptoms that accompany nervousness, such as frequent swallowing, trembling, transpiration, etc., are signs that your body is getting ready for something important. Athletes, stage performers, musicians, and experienced speakers have learned to recognize these symptoms and to appreciate them. They start to worry when these symptoms stay away!

Experience is something that will come in time, by practicing and by evaluating your presentations and those of others. No one in the audience will blame you for being a beginner. However, if you take care to avoid several typical mistakes that beginners (and even experienced speakers) make as discussed above, you will make a very good start with your career as a presenter. If you know and understand the basic principles and you know how to apply these, you are likely to give a talk that is significantly better than the average presentation at international meetings. Hence, lack of experience is not important, provided you prepare your presentation well and you do your best to avoid the obvious mistakes.

In conclusion, giving a successful talk at a conference, a workshop, an instruction course, or just in a work meeting of your own research group all depends on how well you are aware of the two main principles:

- *What is the message I want to convey?*
  *and*
- *How does the audience understand and remember my message best?*

Awareness of how audiences listen and memorize is the key behind a presentation that will be appreciated by many. Learning to apply these two principles is the best recipe for giving effective talks. Feel free to develop your own style, but avoid the usual mistakes so many speakers make.

## 3.5 Posters: ideal enablers of discussion

A successful poster conveys a clear message by means of high-impact visual information and a minimum of text. Posters are very important vehicles for presenting and discussing work at conferences. Poster sessions provide a wonderful forum to meet colleagues and discuss scientific work on a person-to-person basis. Unfortunately, a fairly large number of posters do not succeed in drawing significant attention. Here, we list some of the most frequent mistakes that presenters make and we make some recommendations for composing efficient posters.

### 3.5.1 When is a poster successful?

A poster is successful if it conveys a clear message to the visitor and generates discussion and valuable feedback to the presenter. In order to achieve these goals, the poster needs to be crystal clear about the objectives, the approach, the main results, and the major conclusions of the work, and all this preferably within the proper perspective of existing knowledge on the particular subject.

### 3.5.2 Frequently made mistakes

Too many posters do not succeed in getting their message across. Here are some of the main errors presenters make:

- **Too much text.** Roughly more than half of all posters have way too much text. Posters containing 1000–2000 words (i.e., a complete publication!) are no exception! Who in the audience takes the time to read all this?
- **Unclear structure.** If key elements such as objectives, approach, conclusions, or perspectives are missing, everyone who is not an insider on your subject will not understand why the poster is relevant (and why he/she should spend time on it).
- **Inappropriate structure.** Many people blindly apply the standard structure of a written report (introduction – experimental – results – discussion – conclusion), thereby using their poster as a sort of miniature article, which almost automatically leads to a lot of text. There is no standard structure for a poster, just design a layout that works.
- **Poor figures.** Some figures may be real puzzles, with incomprehensible legends, secret codes, small lettering, cryptic captions, etc. As said earlier, spreadsheet and data programs generally do not produce “reader-friendly” graphics.
- **Too much information.** Many presenters overload their posters with too many data and greatly overestimate the time that the average visitor is willing to spend on the poster.
- **No presenter present.** This is obviously a missed chance for valuable discussion. Another frequent mistake is that presenters take a passive attitude and make no effort to initiate discussions. A presenter who is busy with his/her smartphone does not appear terribly eager to engage in discussion.

### 3.5.3 10 Steps to a successful poster presentation

1) **The message of your poster.** Try to formulate the essence of what you want to present in a single sentence, the message in a sentence. Examples of such sentences are as follows:
   - *I want to convince the audience that my new synthesis method produces a new material of outstanding purity and definition we seek ways for further improvement*
   - *Analyzing reaction X with our advanced and improved micro kinetic model is the key for exploring commercialization of this important process*
   - *The novel ABC technique yields unexpected but beautiful structures of molecules – but we do not understand why these can be stable!*

   The last mini-message might generate some interesting feedback on your poster. Use this sentence as a guide for selecting the data you need to include and the conclusions you want to stress. You probably will not actually print this sentence in the poster, but it helps you to make up your mind what your poster is about.

2) **Introduction.** Write a few brief sentences of introduction (maybe in newspaper header style?) to identify the problem you address, why it is important, what is known about it, the objectives of your work, and your approach to investigate it. Use short sentences and keep this section as concise as possible. Consider if a bulleted list or a graphic might be used?
3) **Results.** Select only the most pertinent results that support your message and remove everything that is not absolutely necessary. Think about attractive ways to present the data in figures. Try to avoid tables as much as possible. Figures and captions should be easy to read – remove all abbreviations and secret codes and use understandable labels only. If possible, add a brief conclusion to every figure, so that there is no misunderstanding what it means (according to you).
4) **Conclusion.** Write the conclusions in short, clear statements, and maybe as a list. Finish with an assessment of what you have achieved in relation to your objectives and, perhaps, what your future plans are.
5) **Attention getters.** What are you going to use to draw people's attention? An attractive title serves as such to some extent but is not enough. Select one of your most important results, a photo or a scheme explaining the scientific background, a model or the main conclusion, or whatever you consider as highlight of your presentation and give it a prominent place on your poster. This is what the audience will see first. It should raise their interest and stimulate them to read your poster.
6) **Layout.** Arrange all the parts of the poster around your attention getter. Add headers if necessary to clarify the structure of your poster, and add everything else that is needed, such as literature and acknowledgements. Ensure that author name(s) and affiliation(s) are on the poster and that you acknowledge who needs to be mentioned.
7) **Review, revise, optimize.** Ask your coauthors and/or colleagues to comment on a draft version of your poster. Assess very critically if the poster indeed conveys the message you want.
8) **Presentation.** Have a "message in 30 seconds" ready for people who walk by and read the title of your poster but have not decided yet if they want to go through the entire story. Prepare a somewhat longer version (1–2 minutes) for those who are interested in the subject. The sooner you get into a question-and-answer mode with the visitor, the more you are likely to benefit from the interaction.
9) **Be proactive.** Prepare yourself for the poster event by studying the list of participants. Are there key scientists whom you would like to discuss the poster with? There is nothing wrong with approaching such people before and asking them kindly to visit the poster because you would be very keen on discussing a certain issue with them. You may for example give your business card, with the poster number written on it.

10) **Follow up.** Consider giving selected visitors an A4-sized reprint of the poster if they show genuine interest. If you already published on the subject, it may be useful to send a PDF in an email with a short "thank you for your interest at the recent conference." (See also the section on contact and relation management in Chapter 4.)

In conclusion, a successful poster conveys a clear message by means of high-impact visual information and a minimum of text. A good poster enables the visitor to grasp the message in a short time, e.g., less than a minute. If the visitor finds the subject of interest, he/she will stay to learn about the details and discuss the work with the presenter. If you fail to get the reader's attention in a short time, he/she is likely to go on to the next poster, unless he/she really wants to know about your work. Search the internet for examples (good or bad, you learn in any case).

## 3.6 Conferences

Conferences, symposia, seminars, workshops, or simply scientific meetings, all of them are ideal events for sharing, exchanging, and getting new ideas for your research, interacting with others, and becoming an active member of your scientific community.

There seem to be ever more conferences every year. "Too many meetings" is the complaint of many academics. Nevertheless, many eagerly engage in organizing new ones or extending that one-time successful meeting into a periodic event. A relatively new phenomenon on the market is commercial congresses, organized by professional but in fact nonscientific organizations, with the simple objective to make money. Whatever, for you, having the organization of a successful scientific meeting, be it a symposium, a workshop, or a summer school, as an achievement on your CV, will certainly be a point in your favor when it comes to promotion or qualifying for a position elsewhere. The purpose of this brief section is to make you aware of the different types of meetings and get you thinking if you could play a role in the organization of such a meeting. We will not go through the entire scenario of planning a conference but refer to some of the excellent resources, in print (Epple 1997) or available on internet (Key; Work Group for Community Health and Development; School of Advanced Study).

### 3.6.1 Types of meetings

Conference and congress are the most general terms for meetings with a program of oral and poster presentations, discussions, and sometimes exhibitions and demonstrations, in a program of plenary and parallel sessions, extending over a few days. These events are often organized by or under the auspices of a scientific organization and held on a periodic basis. A local organizing committee is in charge, but sometimes it follows a general protocol furnished by the parent organization. Examples are the conference series under the European Physical Society, the semi-annual National Meetings of the American Chemical Society, and many others. Such meetings are often visited by several hundreds to thousands of participants and are, in practice, a collection of smaller topical symposia. The large-scale ones offer many attractive features, such as a sizeable exhibition with the latest instrumentation, books, and journals, while the constituent symposia serve as the focal points where sub-communities meet and interact.

Although these large meetings can be overwhelming, they do offer the opportunity to sample what is going on in the field, in addition to the specialized symposia of your own interest. The disadvantage is that it can often be quite difficult to find and meet people and that most participants tend to have busy schedules, trying to follow talks scheduled in different halls of a large convention center (or sometimes even in different hotels). Another problem could be that the program features too many symposia, with some ending up at unfavorable time slots (e.g., Sunday morning, Friday afternoon), or receive less attendance than hoped. As a rule of thumb, such large conventions have peak attendance during the first two or three days, after which attendance tends to fall off.

**Invitations to speak: accept or decline?**

Many scientists receive several invitations per year to speak at scientific meetings, varying from large international congresses to national meetings in their field of research, or educational events such as Summer Schools. Conditions vary from all expenses including travel and accommodation paid, to reduced registration fees, or even no financial advantages at all. Use some of the following criteria to decide:

- Topic(s) of the conference
- Is the organizer a respected scientist on the topic, and is there a respectable advisory board with known scientists?
- Who are the other invited speakers?
- What is the intended audience and is it important that they hear about your work?
- Are you obliged to contribute an article to the proceedings, and if so how will these be published? (many colleagues decline if an article is requested).
- What are the financial conditions (registration, travel, accommodation).
- Does the event fit in your agenda, or does it interrupt other activities too much?
- Can you perhaps send a young coworker instead to gain experience?

In any case, carefully consider if accepting a seemingly flattering invitation is a good idea....

**Workshops** are much smaller than conferences and can have anywhere between 10 and 100 participants. Workshops are usually devoted to one specialized topic, e.g., a specific research methodology, a specific technique, or a recent discovery in the field. The typical format of a one-day workshop is as follows: one or more expert lectures of 30–45 minutes each, several slots of 10–15-minute presentations, ample time for discussion, both plenary and in breaks, room for posters, and an experienced participant who summarizes the discussions at the end of the workshop. Such meetings are relatively easy to organize inside your own university. The costs are usually limited to catering and perhaps some reimbursement of a star speaker, and the fee for participants will be relatively modest. (*Note*: Preregistration and payment of a fee are usually a good idea, as participation free of charge may result in people not showing up at the last minute.)

**Summer schools and discussion meetings** are our favorite type of meetings, as they tend to be more relaxed, with better opportunities for in-depth discussions in a pleasant setting. Often aimed at providing a high-level overview of what is hot in an area, such meetings offer a program with experts who are given ample time to present their recent work, usually in the context of state of a field. Participants can often bring a poster about their work, to present in an evening session (discussion not seldom being facilitated by ample beer and wine). Gordon Research Conferences are a famous series of such meetings in the United States, while Europe has a rich tradition in organizing summer schools aimed at high-level education of PhD students. Often, such meetings take place in the "quiet" rural settings of monasteries, schools in vacation time, or vacation centers outside the tourist season.

A rather unique type of meetings in this category is the Faraday Discussion, organized regularly by the Royal Society of Chemistry of the UK. Here, speakers contribute a manuscript, which is peer reviewed. When admitted to the conference, the papers are distributed to all registered participants well before the conference so that they can read it. During the conference, each contributor summarizes the manuscript in five minutes to fresh up the memories of the audience, after which a lengthy discussion of about half an hour follows. Questions and answers are recorded and published along with the article after the conference. In fact, the collection of discussions on a series of related papers is published as a separate article, with coauthorship of all who submitted a question or a comment. This unique type of meeting gives rise to thorough discussions and exchanges of views, while a large part of the audience is actively involved.

**Minor issues can have major consequences**

In conferences, little details of seemingly minor importance that spontaneously go wrong or have been planned improperly can turn into the talk of the day and negatively affect the atmosphere. We remember a European conference visited by some 1000 participants, where daily lunch was arranged in the form of two sandwiches, snugly packed with an apple, a drink, and a napkin in a paper box – for some an efficient time saver, but a culture shock for many international participants longing for the opportunity to socialize in a relaxed setting over a decent three-course meal with wines. Twenty-five years later, the older part of the community still talks about the "lunch box scandal," but almost no one remembers the excellent program anymore.

### 3.6.2 Opportunities – engage!

Scientific conferences are a marvelous source of information on what is going on in the community, for getting new ideas for your research, meeting your peers, presenting your own work, and getting feedback. But it requires that you, and the coworkers and students you bring along, are active. All too often, we see that conferences are regarded as a sort of school outing, where during coffee breaks, lunches, and other social events, students hang out mostly with their own group – people they already see every day! This is a missed opportunity, of course. Educate them and give them a good example.

Organizing a symposium at a conference, a workshop at your institute, or a summer school is a great opportunity to obtain experience, while you also provide valuable service to the scientific community. Provided you do it well, you raise your visibility and you will certainly make worthwhile new contacts. Many organizations offer the opportunity to propose topics for symposia in their periodic conferences. Maybe you can team up with a somewhat more experienced colleague, make a proposal for a symposium, and hopefully have the chance to organize it – a very valuable experience, and if you do it well, you will be remembered.

Some journals see such specialist symposia as a good occasion to publish a special issue on the topic of the meeting with invited contributions. Why not make a proposal for a thematic issue to such as journal as well? Serving as guest editor is another great opportunity to learn and at the same time raise your profile in the community.

### 3.6.3 Finally, some sources of irritation

Once again, it is not our intention to give a crash course into organizing meetings. Nevertheless, we would like to finish this section by sharing some frustrations from our experience acquired over 40 years in at least 250 scientific meetings of all sorts. Table 3.5 shows a list of personal irritations and things that sometimes go wrong in a conference.

Tab. 3.5: A few issues that sometimes go wrong at conferences from personal experience.

| What sometimes goes wrong... | ...and how it should be |
|---|---|
| **Program** | |
| "Routine" programs reminiscent of last year's conference, same faces as everywhere else on the podium, especially among the invited speakers. | Balanced program of a few high-profile speakers for perspective and (young) people presenting new work. |
| Overloaded with oral contributions, leaving too little time for breaks and poster sessions. | Breaks every 90 minutes at least, adequate attention for poster presentations and discussions; the main value of a conference is networking, not consuming knowledge. |
| Long opening session with irrelevant speeches; sometimes the local guest of honor speaks no English. | Limit opening and closing formalities to the absolute minimum, rather create more time for contributed talks or posters. |
| Parallel symposia not synchronized; people switching sessions miss important parts. | Strict timing is crucial, especially with parallel sessions. |

Tab. 3.5 (continued)

| What sometimes goes wrong... | ...and how it should be |
|---|---|
| **Poster sessions and exhibitions** | |
| Too many posters for too short a time in too little space, sometimes without sufficient light; posters run during lunch and coffee breaks only; presenters are not at their posters. | Take poster sessions seriously, they offer valuable opportunities for discussion and exchange of ideas, as well as networking; keep posters on display for the entire meeting and consider making the posters/exhibition venue the social center of the meeting. |
| Insufficient space or unattractive site for exhibitors; no time in the program for giving proper attention to exhibitors. | Exhibitors often contribute significantly to the budget and help to limit the registration fees; make the event worthwhile for the exhibitors and for the audience. |
| **Organization** | |
| Name badges too small, names cannot be read. | |
| Long queues for registration or for coffee/tea and toilets in the breaks. | Prepare well for logistical details that affect the atmosphere of the meeting; don't surrender too easily to catering services and venue owners. |
| **Venue** | |
| Lecture halls filled up with uncomfortable chairs, sometimes placed too close to each other. | Ensure comfortable seats with enough room; a lay out with table space for every attendant is ideal (though not always possible of course). |
| Noise, sometimes caused by loud conversation outside the halls, by catering services, or by inadequate flexible dividers. | Maybe a member from the organizing committee can be assigned to limit such disturbances to the minimum. |
| Workshops in venues with U-shaped table arrangement, causing stiff necks by having to turn heads towards the front (with views blocked by neighbors). | Audience should have unobstructed views at the podium. |
| **Equipment for presentation and sound system** | |
| Inadequate sound systems, lack of microphones. | |
| Failing presentation equipment, weak laser pointers, or lack of spare batteries. | |
| **Audience** | |
| Mobile phones going off – or even being answered – in the audience; people tweeting, emailing, texting, typing loud; people talking during presentations. | It all seems to indicate that people do not come in the first place to listen and acquire new knowledge, but that meeting face-to-face with others is the most important function of a conference. |

Tab. 3.5 (continued)

| What sometimes goes wrong… | …and how it should be |
|---|---|
| **Chair persons** | |
| Sometimes unprepared, stumble over titles and names, or forgot to make acquaintance with presenters before the session. | Prepare, read abstracts and titles, and check pronunciation while meeting the speakers before the session. |
| Forgetting to enforce timing at the cost of question time or even disarranging the entire program in this way. | Traffic light systems switching to yellow a few minutes before the speaking time is over often work well, red means absolute stop. |
| Allowing questions from seniors only or from the first row only. | Try to engage the audience in the discussion; reserve the first questions for students. |

## References

Dodd, J.S.: The ACS Style Guide; American Chemical Society, Washington, DC, 1997.

Dodd, J.S.; The ACS Style Guide; American Chemical Society, Washington, DC, 1997.

Epple, A.: Organizing Scientific Meetings; Cambridge University Press, Cambridge, UK, 1997.

Garcia, H.E.: The Power of Communication: Skills to Build Trust, Inspire Loyalty, and Lead Effectively; Pearson Education, Upper Saddle River, NJ, 2012.

Key, C.: The Keynote Guide to Planning a Successful Conference; Keynote Networks Ltd. https://spie.org/documents/students/conference_guide.pdf.

School of Advanced Study, University of London: Organizing a Conference, Postgraduate Online Research Training. http://port.modernlanguages.sas.ac.uk.

Smith, R.; Introductions in How to Write a Paper. In Hall, G.M., ed.; BMJ Publishing, London, 1994, pp. 6–15.

Work Group for Community Health and Development at the University of Kansas: Organizing a Conference, The Community Tool Box. http://ctb.ku.edu

Graham Hutchings

# Guest Column: Time, money, and great ideas

*Graham Hutchings is professor of physical chemistry and director of the Cardiff Catalysis Institute at Cardiff University.*

My best advice to young academics setting up their own research program is to focus on doing high-quality research. Choosing the right field and problems to work on is essential. You want to ensure that you are not doing the same as everyone else. Recognize that *great* research takes time, success is not assured, and problem selection is key.

For me, scientific leadership boils down to three key points: great ideas, money, and time.

- Successful research on unique projects has very low odds, you will need lots of ideas to succeed, and hence, you need to constantly generate new ideas. Be ruthless with ideas and select the ones you think are best and will create impact.
- Money: Chase it! To me, there is always the thrill of the hunt. Make sure you are aware of the opportunities and learn how to write convincing proposals; this is an essential skill for a successful researcher. Networking and engaging in collaborations are very good and recommended, but collaborate with your peers, and not with celebrities, as these are likely to get most of the credit. Always make sure that your unique contribution will be visible. In an application, use a track record section to sell yourself, and put in key results that underpin the application.
- Time is the most important commodity that you have; use it wisely, and ensure that you have quality time for new ideas and for thinking deeply about your ongoing projects. Time management is another essential skill; learn it early and apply it. It is essential for being focused, and to avoid that you spread your precious time over lots of outcomes. Involve your team as actively as you can, in writing papers and proposals, submitting abstracts for conferences, and in teaching. They help you, and they learn a lot themselves too.

## Publication strategy: think about a research portfolio

You should also think in an early stage about your publication strategy. If you want to aim for the highest-impact publications, do *not* pre-publish in any way; once you have shared your ideas at a conference, your peers or referees will have a head start. Nevertheless, in the present academic climate, you will have to publish to build and maintain a reputation. The consequence is that you will need a *portfolio* of research, with projects that generate output that can be published rapidly and a smaller fraction of well-chosen high-risk-high-impact research.

### Make sure you and your work are noticed

Try to position yourself and your students for selection as a prize winner. This can be at the beginner's level with the best presentation or poster award or winning a travel grant, or once you are more experienced, you may try to go for prizes for your research or publications. In other words, get yourself and your work noticed.

### Personal Lessons

I end with a few lessons learnt in industry and academia.

- Recognize when you are lucky – some of the most important discoveries are made when an unexpected result is obtained that challenges rather than confirms the hypothesis being tested. Remember "chance favours only the prepared mind" (Pasteur).
- Learn to manage your manager, and if you don't like the system, change it! This was advice I was given by my first boss when I joined ICI (Imperial Chemical Industries) – mind you, he was both very successful and unmanageable.
- How do you make sure you get the time to make that great breakthrough? Once again, time management is key.
- In your team, ensure that it is clear "Who has the monkey" – I learnt this in a management course while in AECI (African Explosives and Chemical Industries). Think of a task as a monkey that sits on the back of the person who has to get the job done. Make sure that you don't allow the upward delegation of problems and partly completed assignments, because then, you will end up with all the monkeys' back with you on a Friday afternoon. If this happens, it can cause you a lot of your valuable time to be wasted – don't allow it. In your team, educate your people to be problem solvers, and not problem creators. So if you are managing a team, make sure you control the workflow to and from you!
- Leadership is about seeing all sides and not asking your team to do something that you yourself are not prepared to do.

Finally, take a step back now and then to reflect a little on what your aims are. Are you heading for a future in academia or in industry? Are your work and personal life in balance? Considering these things is important early on. Overall, remember that research is fun and being the leader of a research team is a great privilege. Finding out something new is exciting and can lead to great rewards for both you and society. But if it isn't fun, it's not worth the effort in my view.

# 4 Management skills for researchers

## 4.1 Introduction: the scientist as manager

As we described in Chapters 1 and 2, being a successful scientist requires activities on several fronts. In his Guest Column (page 71), Graham Hutchings presents his three most important enablers for high quality research: "great ideas, money, and time." Running a large laboratory that covers several research topics obviously needs a smooth organization to ensure that each subject receives the attention it deserves and the resources it needs to become a success. But the same is true for the much smaller research activity of a young assistant professor or a junior research scientist. Investing in skills that help you to organize your work efficiently and effectively is important, and it is wise to learn at least some of these in the beginning of your career. We therefore devote this chapter to management skills that we believe are particularly helpful for a researcher. You probably don't have to become an expert in the finer details of all these skills, but knowing the principles so that you can regularly think about how these can help you in your work is what we are aiming for.

## 4.2 Fundraising: who will pay your research?

Research is expensive. You need a laboratory with specific infrastructure. The chemist will need a well-ventilated lab with fume hoods, workbenches with electricity, certain gases, flowing hot and cold water, lots of glassware, analytical instruments, safety installations, storage cabinets for chemicals and gas bottles, computers and data lines, etc., while an engineer likely wants a laboratory where he/she can design and build models, demo systems, pilot plants, perhaps supported by a small workshop. A theoretician needs computers, space to put them, and ultra-fast data connections to super computers or personal computer clusters, and all need offices and desks for people to work on. You most likely also benefit from central services, which need to be paid by somebody. If you are lucky, your university provides much of this, but many times, you yourself will have to arrange the funds to support students and postdocs, to buy specific instrumentation and consumables, and to travel to conferences.

### 4.2.1 How are universities and research groups funded?

It is useful to have some insight in the way universities are financed and how your own research group fits within this. Although variations across different places in the world and even across institutions within one country are large, the scheme in Fig. 4.1 probably captures the essence of how a university is funded.

https://doi.org/10.1515/9783110468892-004

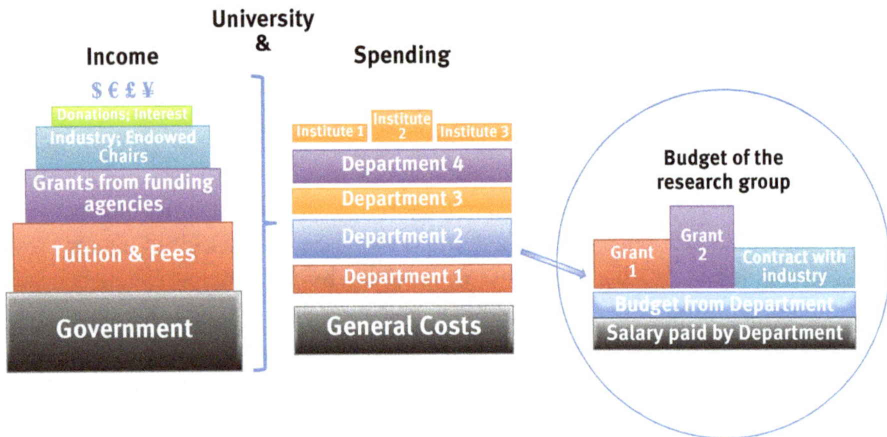


Fig. 4.1: Funding structure of a university; please note that variations between different places in the world may be large.

Most academic institutions receive an annual budget from the local or national government. On top of this, they charge tuition fees, although there are also countries where education is free and covered by the government. Then there may be income from funding organizations, industry, donations from alumni or philanthropic institutions, legacies, and interest from university capital and dividend from shares. To give an order of magnitude estimate, a typical mid-size university in Europe would have an annual budget of a few hundred million euro, while a top university in the United States might be good for $1.5 billion income per year.

Each year, the executive management of the university (i.e., the Rector Magnificus, the provost, the president, or perhaps the governing board) will determine how the money is divided over central costs (administration, buildings, central services, etc.) and the departments and/or institutes. In a department, the dean or head of department is usually the one who decides over how the budget is spent. Maybe he/she uses transparent schemes for allocating money or he/she does it to his/her own wisdom and discretion. Again, it is important to realize that budget models can vary widely, which will impact the budget of your research group and the freedom you have to spend it. We give a few examples on how a department can be funded for reference and recommend that you find out yourself how it works in your environment.

Example 1: At mid-sized university A, department X receives its income from the central budget provided by the university plus the 15% overhead it charges on all external funds brought in by the staff. The department pays the salaries of all permanent staff (tenured academics, tech and admin support). All full professors are in the fortunate position to receive two university-paid PhD students. The central budget covers the costs of teaching, administration, maintenance and use of the building and all utilities, the workshop, the library with journals and books, and the cost of some large shared research infrastructure (such as an electron microscope, a computer cluster). If research groups receive external grants, they pay 15–25% overhead to the department; the rest is used for the stipend or salary costs of their non-tenured coworkers and running budget for consumables, travel, and perhaps some investments. Newly appointed academic staff usually negotiates with the dean that they receive start-up funds, for equipment, extra PhD students, or a postdoc.

In a scheme like this, which is generally common in Europe, the tenured professors have job security and freedom to determine their own research subjects, as long they teach well and manage to get research done. However, to run a substantial research program, they will have to be successful at bringing in funds from agencies such as the National Science Foundation or from bilateral contracts with industries.

Example 2: Department B at another university operates in a way that professors are forced to be real entrepreneurs. The central budget pays for the buildings, central administration services, and teaching, but the rest of the budget is distributed among the academic staff, according to a distribution scheme that scales with output performance (e.g., number of diplomas, publications, patents, and credit points for teaching) and the generation of funds for their respective group. The professors have freedom to spend the budget on salaries, including their own, equipment, consumables, and travel, but they must acquire grants and contracts to be able to work and secure their own job in the next years.

The positive effect of this system is that success is rewarded by a larger share from the central budget, which works as an incentive for success. The downside is that tenure for the professors is no longer a guarantee for job security and that the capacity to bring in funds is more important than academic excellence. Risky situations may arise if too many groups become too successful, and the departmental funds for each individual group decrease, or if the department becomes overly optimistic in successful times and appoints more academic staff than the system can sustain when the funding climate becomes less favorable. In such a case, the department may find itself forced to reorganize and lay off people. We know examples where funding mechanisms of this type have led to situations where the university is not much more than a "hotel," which offers space and some minimal support, while professors are supposed to generate income needed for all their activities, including the teaching.

As said, the two examples represent extreme cases, and several other funding schemes exist. It may take some time before a young starting professor understands how it works in his/her environment. We strongly recommend figuring this out as early as possible, for example, at the time of your job interview.

It is quite normal that candidates who are offered a position negotiate with the university to obtain start funding, before they sign the contract. If the university is eager to have you, they are probably willing to help you with some investments and perhaps some PhD students and one or more postdocs. Although the space for negotiation tends to be limited these days, you should always bring this up during your application process (and not after you have signed a contract).

### 4.2.2 How is your research group funded?

The next question is how much base funding the university or, rather, your department will give you each year. Will they just pay your salary (some universities only pay you during the teaching term, i.e., for 9 months), and can you use the laboratory space for free, or are you supposed to share in the costs for space, utilities, and overhead? Can you make use of central equipment or facilities of other groups, and do you have to pay for this? Do you receive a basic budget for consumables, or will you have to find funding for all these costs? Who pays the stipends or salaries of PhD students? Conditions may vary widely over the world.

Inevitably, you will want to go after external funding at some stage in your career, because you need the money for research as well as because you want it on your curriculum vitae.

### 4.2.3 Writing an effective grant proposal

We assume you have a viable plan for an excellent research project, which fits with your profile. This may sound trivial, but too many scientists do the opposite: they look for funding opportunities and then start thinking what might be a good proposal. Consider that writing a convincing grant proposal takes a lot of time. Do you want to invest this time, or is it better to spend the time for writing up your research in articles first, to give you more credibility and a stronger position to apply next year?

Orient yourself on the types of funding sources available. Special grants for young researchers may be smaller but are a good step up for larger projects in the future. Some countries have several sources where you can apply. For example, National Science Foundations tend to have funding programs for basic or fundamental research, while there may be separate funds for applied projects together with industry. Carefully read the instructions, and make sure you understand the goals of the funding program you are applying to. Proposals aiming at testing a clearly formulated hypothesis tend to do better in fundamental science programs, while explorative journeys into the unknown but with big reward if the idea works are better suited for innovative programs. It is a good idea to discuss your idea for a proposal with a program officer, these people are generally keen on funding interesting ideas of promising newcomers, and they are usually very open to discuss their programs with you.

Networking and belonging to the community in which you plan to do research is vital (see also the section on contact and relation management later in this chapter). It is a must to know the scientists of the field where you apply, and they should know you. So, be present and speak at their national meetings, or visit some of these groups, and make sure that your peers as well as the leaders in your area know about your work and your ideas. Maybe there are valuable opportunities for collaboration. Anyway, some of the more experienced members of the community are likely to be the reviewers of your proposal, so it is important for you that they have a good impression of your potential. And of course, it may be a good idea to cite their work if applicable, both in papers and grant applications.

Write the proposal around a clear idea, and keep this as the focus. The "why" of your idea is as important as "what" and "how." Everything we wrote about the importance of a clear message in presenting science (Chapter 3) is as important here as in a publication, a poster, or a talk. Avoid long technical diversions, and if these are absolutely needed, consider putting these in an appendix or a text box, so that they do not interrupt the main story. This way, reviewers will notice that you know your subject, but they do not have to read it all if they do not want it. Be sure to cite pertinent papers and the gaps in understanding that you address.

Make sure the proposal is clear for scientists in the broader discipline, but do not assume that they are necessarily an expert in your specialization. So, avoid typical jargon, abbreviations, figures and tables for the insider, etc., and explain also how your expertise fits within the broader discipline and indicate what the state of the art in your field is. Well-laid-out schemes, figures with clear captions, and other attractive elements to break long pages of text may let your proposal stand out among the competition.

**15 minutes per proposal...**

As a reviewer of a series of proposals, I (JWN) typically spend 10–15 minutes to read the proposal. If I do not understand what is proposed, I give up and give a negative recommendation. If the aims are clear, I am happy to invest more time and provide detailed comments for the program jury to base a fair decision on and for the submitter to clarify and improve the proposal, if necessary.

Having the essence of the proposal in terms of key ideas, criteria for success, and requested resources outlined in the first page(s) will help reviewers a lot in their evaluation. Just keep in mind that reviewers are busy people. If the review is organized in panel sessions, the members may have to work themselves through 10–30 of these proposals in a day. Concise and crystal-clear applications will certainly be appreciated and complicated puzzles are likely to be put aside.

Potential risks and pitfalls, or controversies in the field, should be mentioned and assessed, and having an alternative approach if the original one should fail may help to give the reviewers more confidence in you.

Avoid asking for unreasonable items. For example, it may be an excellent idea to request budget for a one-day-per-week technician or for workshop hours, but don't overdo it, and justify why you need such extra resources.

Having an experienced professor as mentor, who is willing to listen to your ideas and read your draft proposal in an early stage, can be very valuable. Such a mentor does not have to come from your own department at all. With some distance from your subject, he/she may very well comment on how effectively you explain the essence of your ideas. Feedback is vitally important for you.

Once you have the proposal ready, let it rest for a few days (if deadlines allow). When you have read it again, decide if it has chance to score high. Don't gamble with a mediocre or incomplete proposal, it may hurt your reputation at the funding agency and with the reviewers, who will probably know you!

Depending on the procedures, you may get the chance to reply to the reviewers and make minor adjustments before the grant organization takes its final decisions. This is a chance to correct misunderstandings or wrongly interpreted parts or provide additional explanation. Our experience is that reviewers who know the field are generally fair and constructive, but occasionally, we have also received some nasty reports of people who obviously did not believe in our ideas (or in us). Get angry as much as you like, but when you reply, you are courteous and to the point, and you explain concisely but clearly why the reviewer is wrong. Keep it short, because long replies to referee comments don't go well with juries. If there is too much to correct, there was something wrong, they will probably argue. If you sense unreasonable hostility or evident lack of expertise in a report, then why not call the program officer and ask advice on how to deal with the unjustified criticism?

The good old days when we entered the academic world, where more than half of the proposals were granted, are unfortunately over. Typical success rates have dropped tremendously, sometimes below 5%. The European Union often works with a grade system, where your proposal has to score at least 80% to be eligible for funding. If you end above the threshold (always a happy moment to hear), it does not mean that you will get the proposal granted, unfortunately. Nevertheless, you should be encouraged by the outcome. Resubmission of an application that was close to success should always be considered seriously. Maybe you can discuss the outcome with the program officer and do your utmost to improve the proposal for the next round.

## 4.3 Time management – creating quality time to think

Having enough quality time to think, to be creative, and to do important things, whatever these are, is something we always seem to be short of. It is an intriguing dilemma that everyone of us disposes over all the time, 24 hours per day, 7 days per week, that is available, and nevertheless, we always seem to have too little of it.

Time management is about being effective and making the best of the time that you have. There are many books that treat the subject and provide simple or quite complicated schemes to help planning and keeping track of activities (Hindle 1998), but we limit ourselves here to a few basic and rather practical principles that have helped the authors to manage their time.

Intermezzo: Tips for busy people

*Doing the right things at the right time*

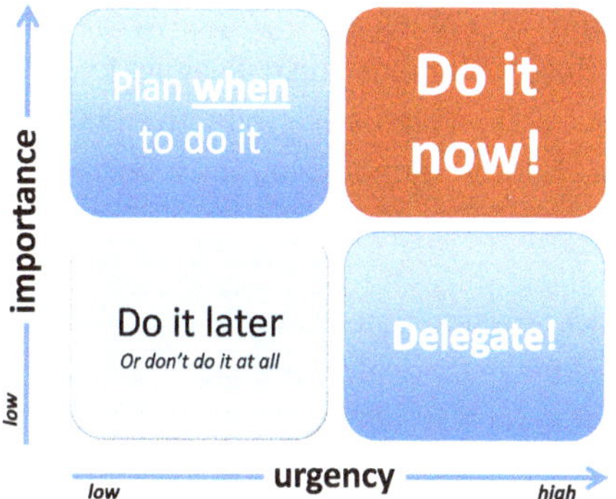


Fig. 4.2: Time management is about doing the right things at the right time. The scheme is also known as the Eisenhower model of time management.

### 4.3.1 Priorities: urgent is not the same as important

The first point to realize is that looking at the many tasks and activities we are supposed to do, some are important while others are urgent, see Fig. 4.2. However, importance and urgency are not the same. Cleaning the laboratory may be urgently needed, because the dean or director will come around with important visitors tomorrow. However, for you finishing an important grant application before the deadline expires may have much higher priority than meeting the dean's guests; in other words, this task is both important and urgent. Another example: you need to prepare well ahead of time for a prestigious keynote address in three months from now. This is of great importance, but not immediately urgent as you might also choose to start preparations next weekend.

Hence, our first recommendation is to keep a list of tasks and activities and include priority, urgency, and deadline. Many email/calendar systems enable you to make such a list, along with timely reminders, and indications how far you already progressed. Ensure that for each task, the purpose is clear. Make it a habit to start the day by looking at the list and by ending the day to update it.

### 4.3.2 Personal quality time zones during the day

When one of the authors (JWN) became an associate professor in 1989 and had several PhD students, he also worked on a textbook. The other author (JKF) made him aware that certain activities are best done during certain parts of the day and that it is perfectly fine to declare such time zones off limits for others. In this way, JWN found out that for him, the early morning hours were best suited for creative activity, i.e., writing, designing figures, studying literature, and thinking deeply about content. Hence, as far as possible, the time until the coffee break at 10:30 am was "official quality time," and the office door remained closed. Students and colleagues knew the pattern and generally did not disturb before coffee. A similarly productive time zone arose in the evening, after dinner, when the children were in bed. The afternoon, on the other hand, was better suited for meetings or for typical managerial duties and administrative work.

Discovering your own patterns of effectiveness and time zones for specific activities and enforcing these can be a great way to make more of your time.

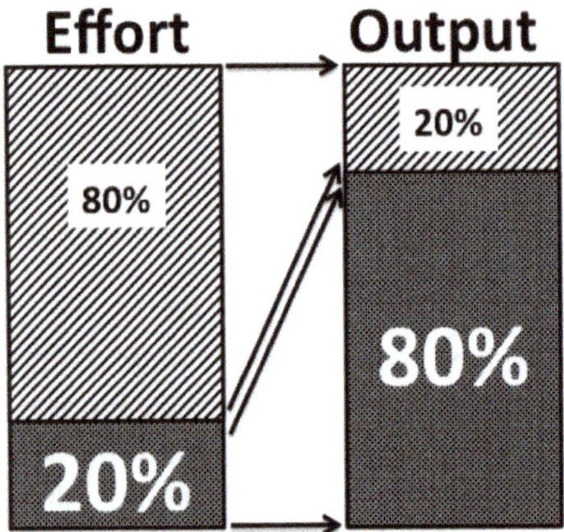


Fig. 4.3: According to Pareto's principle, you produce 80% of your useful output with 20% of your effort or, in other words, 80% of your production during 20% of your time.

### 4.3.3 The Pareto Principle: the 80/20 rule

It is a comforting and reassuring idea that output is not necessarily linearly related to the effort spent. Applying Vilfredo Pareto's 80/20 rule to time management suggests that 80% of your useful production results from only 20% of your time, see Fig. 4.3. There are many ways to look at this, but our experience tells there is a lot of truth in Pareto's principle. Many researchers will probably agree that from the time spent on a specific project, including all the efforts on building up, learning how to do things, and trying out ideas, the definitive results that ended up in the final report or scientific publication accounted for only a fraction of the time. Of course, it is not possible to have the top of the iceberg without the large part that is under water (although policy makers sometimes seem to forget this), and in hindsight, one could always have reached the goal faster than factually happened.

But the important message of the Pareto Principle is that for familiar types of work, that is, in which one has a certain experience, a substantial amount of useful output can be generated in a relatively small amount of time. Giving priority to such elements of the job, for example, in realizing deliverables requested by your superiors, having your teaching courses for the students prepared in time, or performing a routine project for an external customer to generate extra income for your group, ensures guaranteed output and ensures that you can afford to set aside quality time for the more challenging parts of your work.

### 4.3.4 Avoid wasting time

Of course, many external factors can disrupt your ideal work patterns entirely or use up disproportional amounts of time:

- Email and messages from social networks, especially if your mobile phone produces a sound each time a new message comes in. Our radical solution is to ignore non-professional social networks completely and pay minimal attention to research networks such as ResearchGate or Mendeley. Principally, these offer useful services and help to spread your work among the community; however, unfortunately, they regularly call for attention, sometimes several times per day. Emails, we try to check only a few times per day, we switch off notification services that beep every time a message comes in, and we often leave our phone on our desk when we go to meetings or seminars.
- WhatsApp, WeChat, text messaging – all are great communication tools, but keep them under control; i.e., do not allow them to disturb. For example, tell colleagues and students that they cannot rely on these message services to inform you or reach you. If urgent, use a phone, or if it can wait, use email.
- Colleagues, students, and sales people who come by your office and expect your instantaneous availability: you do not want them to interrupt your periods of quality time. The open-door policy can ruin your productivity, although on the other hand, you also want to be regularly available for your people. Practicing the open-door policy only during certain times of the day is always a workable solution for us.
- Meetings, and especially periodical ones. The typical once-per-week management team meeting can be useful to ensure that all members are on the same page with respect to the organization but can also become a time waster when insufficiently prepared or when a standard agenda invites too much pondering on issues that do not need to be dealt with every time. As a rule, routine meetings where participants come unprepared are inefficient. Consider what needs to be discussed in the team and what is better handled face to face with individuals.

These remarks about wasting time at meetings are not meant to say that all meetings are always useless. On the contrary, they can be great sources of inspiration, of chances to listen to others, to exchange ideas, to get people united behind common goals, and to take jointly supported decisions. Limiting them in duration, however, is usually a good idea, and therefore, having a clock visible to all in any meeting room, as well as in your office, is a must. Just a 15-minute coffee meeting to start the week and to inform each other on the actual issues and priorities can work very well; at the same time, it is good for the social atmosphere and cohesion in your group.

### 4.3.5 Time management of coworkers: delegate small projects instead of tasks

An often hearsay frustration is that managers always seem to run out of time, while their subordinates run out of things to do (Oncken & Wass 1974). For academics, this situation hardly occurs in the professor-student relation and more general in the relation between professionals who have their own areas of responsibility and associated span of control. It can play a role in the relationship with technical or admin support staff. If you recognize this in your own environment, then consider whether your way of delegating tasks is optimal. It makes a difference whether you give assignments in concrete form of a to-do list (first do this, then that, report back when done, etc.) or that you delegate work as a project along with the purpose, the desired end-result, and the appropriate responsibility to do what is needed. Completing such a project will generally give much more satisfaction to your coworker, than ticking off small tasks from a to-do list.

#### 4.3.5.1 Briefing properly is key

When it comes to delegating work to your coworkers, support staff, and students, giving them the appropriate information is of course essential. Briefing is key, and it takes time. Much confusion and misunderstanding can result if this phase is underestimated.

- If you delegate work to new or inexperienced persons, always brief them in a personal meeting and monitor progress and keep in touch at least daily. Remember that praise is always more effective than criticism, and starting off with praise gives better chances that the other will be open to critical feedback as well.
- If, however, work is delegated to an experienced coworker, briefing can be done in writing (email, memo), but personal contact is always preferable. Monitor progress as and when needed and agree beforehand on how to keep contact on the project.

In delegating, it is essential to explain clearly the objectives of the work and why it is important. If persuasion is needed, facts and context work better than emotional arguments. Also consider the other's priorities and personal goals and whether the assigned work is a good match.

An important principle of delegating work is that one never should ask others something to do that oneself would not be prepared to do; e.g., resolving a conflict at a higher level is something you should always do yourself. If it involves a coworker, one should solve it together, with the one who delegates – you – in the lead.

#### 4.3.5.2 Upward delegation

Be careful in avoiding that your people come back with partly finished assignments, hoping that you will do the final part. If that is not what you agreed at the start, then do not accept incomplete work. It may be tempting for you to give in, as it may take you only little time to apply the finishing touch, but the danger is that next time they try it again and return the job in an even earlier stage. Our recommendation is to be very strict: Once a task has been assigned along with proper briefing and transfer of responsibilities, it is your coworker's responsibility to complete it. In case of unforeseen difficulties or circumstances, you will be available for advice, but completion of the task should not be delegated back to you.

### 4.3.6 Cultural aspects in the perception of time and the way it is spent

In the international setting of scientific research, one must be aware of the cultural background of team members. Significant differences with respect to the concept of time and the way it is spent can exist. Some cultures place much value on spending long hours on the job, while in other countries, the emphasis is more on balance between time spent on work and on private life and spending work time as effectively as possible. Working long hours is the norm in the United States and in Asia, while in Europe, people focus more on a healthy work-life balance. Which approach leads to more success, on short- and long-term? We find it hard to say, although we know what we prefer.

We are personally strongly in favor of allowing our coworkers flexible work hours, following the principle of working when it is most effective. For example, when a certain research facility in high demand is available, or a specific instrument is in top form, a motivated scientist will want to use it, no matter when it is after regular work hours (provided safety is guaranteed of course). Enforcing strict work hours for motivated professionals is unnecessary. However, not all work cultures allow such flexibility.

Another cultural difference in this respect is the importance of being on time for appointments or obeying the time schedules of events. For people from the West, flexibility and improvisation, with respect to appointments in other parts of the world, can at first sight be disturbing and unsettling. However, our experience is that with a bit of determination and persuasion to stick to important items in a program, in the end, one almost always leaves with having the planned visits and meetings accomplished, be it in a different order or timing than initially foreseen.

#### 4.3.6.1 Messy desks: effective or not?

Finally, time-management literature generally stresses the importance of clean desks, simple file systems in your computer and in your file cabinet, and avoiding piles of documents and articles at the work table and random files at your computer desktop. We are not so convinced that this is always relevant. Obviously, one takes no risk with measurement outcomes and important data, even so with any document regarding financial administration, such as evidence of purchases and trips required for reimbursement from your own institution, or for being available when requested by grant agencies. We use a simple, systematic file system for anything that may be requested in the future, to avoid financial penalties or missed chances. However, we tend to be less systematic in filing articles, literature, and notes. Sometimes, these do pile up on our desks, amidst notebooks and other stuff. "A messy desk is a sign of genius," is the saying, or according to Albert Einstein, "If a cluttered desk is a sign of a cluttered mind, of what, then, is an empty desk a sign?" A recent study (Vohs et al. 2013) titled "Physical order produces healthy choices, generosity and conventionality, whereas disorder produces creativity" gives confidence for not being overly concerned with the state of our desks.

## 4.4 A little project management

Scientists do not like project management. They associate it with activities that can be planned, for which the outcome is known from the start, and with intermediate results that must be delivered at specific times ("milestones"). Indeed, in this sense, research as a venture into the unknown, with uncertain outcomes that can only be hoped for and often are entirely different than expected, is hard to approach as a project. Nevertheless, many funding agencies require that proposals are written as a project, along with some form of planning. Ignoring this would mean we would lose a chance of funding.

However, many activities in academic or industrial research environments are suited for planning and can very well be treated as projects. Just think about building a new laboratory, a new experimental setup, writing a publication, preparing a presentation for a conference, developing a new course or an entirely new curriculum, organizing a student excursion, a workshop or a conference, completing the final phase of a PhD thesis, etc. Here, project management can be of great value to accomplish the tasks in time.

Project management as a discipline is highly developed. Taking a formal course can easily go to deep for your needs, as it usually caters for more complex situations and thus spends ample time on the administrative aspects. Hence, we believe it is useful to present here a short summary of project management, as the "minimum you want to know." We assume that you yourself will be the principal as well as the client who will use the outcome of the project.

### 4.4.1 Definitions: projects differ from routine work

Let us start with proper definitions of the terms project and project management. A project is a time-limited activity for achieving a unique and tangible end-result (a product, a service), often defined in the form of deliverables along with specifications in terms of quality. The project has a defined beginning and end and usually a fixed budget. Note that the words "unique" and "time limited" imply that a project is a special, one-off activity, has new content, and can never be a routine job. The management of such a project is therefore different from that of normal day-today activities. Project management encompasses the initiation, planning, organization, control, and monitoring of all the tasks and resources in the execution, required to achieve the intended result of the project, including the coordination of a project team, in case others are helping you.[1]

### 4.4.2 Scope, time, and resources

Three types of resources constrain your project: the **scope** (how broad will you allow the project to be, which aspects are included and which not), the **time** in which it must be completed, and the **resources** (or budget) that are available, in terms of money and work hours of people in the project team or external service providers. Cost is not always fully visible in academic environments, as time spent by academic personnel and support staff is often not accounted for (and we do not advocate that it should).

Defining the scope of a project for research, characterized by an unknown outcome, is not straightforward at all. Between the time while formulating a research proposal and the moment a research starts, the scope often must be adjusted, for example, because new knowledge has become available, or a new instrument has just been introduced on the market, or you met a new colleague who can bring in unexpected but welcome expertise. Adjustment of scope will probably be needed several times during the project, with implications for timing and budget. Sometimes, this phenomenon is referred to as "scope creep," giving it a negative connotation, but in research, this is just a fact of life.

1 https://wiki.en.it-processmaps.com/

EXAMPLE: The dean has instructed you to develop a short hands-on course on computer skills for first year students (typical cohorts consist of 150 students) to consult the library, learn to use referencing software, and use graphical software packages for statistical data analysis. He wants it to be implemented the first semester of the next study year; he makes 160 hours of IT support available and provides a budget for software and hardware of €25,000.

Project Scope: Introductory computer skills on..., for 150 students, to be completed in 12 study hours
Time: Deadline for implementation, September 1 next year.
Money: Budget of €25,000, 160 man hours of IT support, your own time.

### 4.4.3 Stages in a project

The discipline of project management disposes over several different approaches and methodologies for various situations. We limit ourselves here to relatively simple straightforward projects, with stages as shown in the diagram of Fig. 4.4: initiation, planning, execution, continuous monitoring and control, and closing.

Fig. 4.4: The stages of a project; note the planning-execution-control cycle, which can be traveled through for several rounds and gives the project an iterative nature.

**Initiation:** The project and its intended outcome are defined. A detailed analysis of the existing and the new situation should be available, and the reasons for the desired change or for reaching the intended result should be crystal clear. Define the objective and scope of the project, the deliverables, the specifications, the costs, and the moment the project should be ready. Check if the available budget is sufficient. A SWOT analysis as discussed in Chapter 5 may be appropriate to assess if the goals are realistic and the project is viable. The selection of a project team forms also part of this stage, and for a very large project, possibly a steering committee is required. The initiation or definition phase of the project is very important and should lead to a clear and detailed project plan. Underestimating this stage may later on lead to serious problems and misunderstandings.

Planning: A detailed scheme is made of what needs to be done at what time; who has which responsibility; what are the costs in terms of budget, time, manpower and other resources; and an organization and governance structure including a communication plan (project meetings, how to report progress, and how to share information). A kick-off meeting is scheduled to make sure that all team members and other stakeholders understand the project, its reason, and the way it will be executed.

The obvious tool for planning the typical small-to-medium-sized projects in the academic world is the Gantt chart, a tool developed already in the early 20th century, by Henry Gantt. In its simplest form, it is a bar chart with time on the horizontal axis, indicating which phase of a project is executed at what time (see Fig. 4.5). More elaborate versions indicate the milestones and the interdependencies of subtasks, e.g., by colors or arrows. It is easily created with the help of a spreadsheet program, and many templates are available on the internet.

Execution: The project team carries out the various activities in accordance with the plan.

PROJECT X

| | | | | | | | | | | |
|---|---|---|---|---|---|---|---|---|---|---|
| task 1 | | | | | milestone A | | | | | |
| task 2 | | | | | | | | | | |
| task 3 | | | | | | | milestone B | | | |
| task 4 | | | | | | | | | milestone C | |
| task 5 | | | | | | | | | | |
| task 6 | | | | | | | | | | |
| months | 1 | 2 | 3 | 4 | 5 | 6 | 7 | 8 | 9 | 10 |

Fig. 4.5: The Gantt Chart: simple but useful tool to visualize the planning of a project.

Monitoring and controlling: The almost continuous cycle of measuring the progress of the project, evaluating where it is supposed to be, deciding on how the activities and the plan should be corrected, along with keeping track of costs, hours spent, and adjustment of the plans. It is about keeping the project on track, on time, and within budget.

Some grant agents request an audit upon completion of a project. Not having proper documentation on all financial matters or hours spent can involve penalties, implying that one does not obtain the entire budget. If you can afford a – perhaps part-time – trustworthy administrative assistant, you may be able to save a lot of time and frustration.

The three phases of execution, monitoring/control, and planning form a cycle together through which the project iterates several times.

**Closure:** The final phase of a project implies ensuring that all project goals have been realized and the formal acceptance of the outcome by the principal. This phase also includes a report (administration, costs, bookkeeping, time sheets, logs, etc.) and an evaluation with the team to learn and to celebrate. Maybe it is a nice idea to convert the Gantt Chart into a historical timeline, for example, by using the most important phases and achievements, as a graphical account of how the project was carried out. It may help you and others in future projects, and it can be a nice wall chart in your laboratory for showing visitors.

### 4.4.4 Scientists prefer agile management styles

As mentioned already a few times, researchers, and certainly those in academic environments, typically dislike the formal organization of project management. They are not the only branch. Since the turn of the century, agile management has come into fashion. This (project) management style is based on the Agile Manifesto (Beck et al. 2001), which highlights four values:

- Communication with stakeholders is more important than obeying standard procedures.
- Focus on delivering products and services that work and satisfy the needs of the customer, and less on providing legally sound, and often voluminous manuals to document the outcome.
- Close collaboration with the client in all phases of the project.
- Openness to changes in scope during the project.

Such ideas are generally closer to the heart of a scientist, who is used to progressing insights and changing scopes versus budgets and resources that are typically constrained. Nevertheless, we do not propose to abandon all principles of project management, but we do like the flexibility and iterative possibilities of the agile approach.

## 4.5 Knowledge management and capturing implicit know-how

Researchers, educators, scientists, developers, all of them deal with information, data, insights – for simplicity, and probably not entirely correct, we catch it all under the term "knowledge." Organizing literature is conveniently done these days with so many sources on line and tools to build your personal databases and reference managers. Cloud services also greatly facilitate the storage of data and documents in a way that your entire team can access these and contribute. Not all information can be treated in this way, however. The management of knowledge in organizations has become an important discipline since the 1990s, and several multinational companies now have a CKO (chief knowledge officer) in addition to the CTO (chief technology officer) for technology and R&D (Koenig 2012).

Davenport's initial definition (Davenport 1994) states it succinctly: "Knowledge management is the process of capturing, distributing, and effectively using knowledge." But there is more to it.

Knowledge comes essentially in three forms: ***explicit*** (knowledge set out in tangible form, i.e., which has been documented), ***implicit*** (knowledge in people's mind, which, however, can be made explicit), and ***tacit*** (knowledge in the form of experience or intuition that is almost impossible to document). An example of tacit knowledge is the "Fingerspitzengefühl" of a restaurant chef who knows exactly how to prepare his/her signature dish but would not be able to write it down exactly in a way that somebody else could reproduce it (and, probably, he/she would not want it).

Recognition that knowledge comes in different forms is a fact in the nowadays generally accepted definition (Duhon 1998): "Knowledge management is a discipline that promotes an integrated approach to identifying, capturing, evaluating, retrieving, and sharing all of an enterprise's information assets. These assets may include databases, documents, policies, procedures, and previously uncaptured expertise and experience in individual workers."

We owe this comprehensive definition of knowledge management to Duhon, who, in 1998, published an article under the all-revealing title "It's all in our heads." That is exactly where a big concern of research leaders lies: How do we capture the knowledge of students and postdocs who worked with us for a few years when they leave? Publications and theses usually emphasize successful endeavors. Laboratory notebooks – provided your people fill these with the discipline that may be expected from professionals! – can give valuable extra information, and all data collected have probably been stored in a comprehensible and retrievable way, maybe even the PDFs of articles referenced in theses and publications. But much of the know-how leaves the laboratory along with people in the form of tacit knowledge. It is important to realize this at an early stage. Succession planning can prevent the loss of implicit knowledge to some extent, when successors overlap in time with the people they succeed. Permanent technicians in the laboratory are also a great asset in this respect. Large research groups have better opportunities for such policies than smaller ones do, and the funding structure of universities or funding agencies does not generally allow effective succession planning.

### 4.5.1 Document essential procedures and "lessons learned"

To prevent essential knowledge and know-how from leaving your laboratory, think about implementing a strict policy for your students and coworkers to carefully document essential procedures, by writing a step-by-step manual on how to use an instrument, to carry out a specific synthesis, or to collect and analyze data. Such homemade manuals should be regularly updated, for example, with information on how to replace parts in a machine, how to solve specific problems, tricks in using the software, etc. Our strong advice is to do this from the installation and first use onward, even if the vendor provided a good manual. Such documentation is of tremendous value for new users, and it is an effective way to safeguard essential know-how. It is also good practice to capture essential know-how and knowledge at the end of each project, by organizing a "lessons-learned" meeting with all people involved.

## 4.6 Contact and relation management

Keeping track of important contacts is important for researchers and academics. We may want to remember not only our current students and alumni, past coworkers, and the peers with whom we interacted or may do so in the future but also vendors, sales representatives, service engineers for your infrastructure, publishers, editors, funding agents, administrators, and science journalists, whom you may need to contact again in the future. And many people will have moved on to new positions after a few years.

Hence, it is good to have a system that captures names and contact info, along with your notes on what the reason was to have contact.

The old way was to collect business cards and make notes on them and keep them in a card index box or rolodex system; but nowadays, there are many software tools to help you. These range from the standard contacts function of your email system to programs such as Evernote, which let you digitize business cards and automatically complete the information by searching in online social platforms such as LinkedIn or Facebook.

Maybe you do not need this. Maybe you are the person who remembers every important meeting and conversation at a meeting that you had. We, however, do not (maybe it is our age ☺), and therefore, we rely on simple forms of contact management, and we recommend you regarding to adopt a system for yourself at an early stage.

### 4.6.1 Relation management

One step higher than keeping track of contacts is what you do with them: relation management. Are you making use of your contacts and are you good at building mutually beneficial relationships in a network of peers, professionals, former students, friends, etc.? Do you keep the most important among them updated about your projects and progress? When you published an important article, do you send PDFs to selected contacts with a short but personalized message? Again, having a consistent and adequate presence on social platforms can effectuate this to some extent. Some of our colleagues regularly publish a brief newsletter, for example, at the end of each year, and use it as Christmas greetings, via email or regular mail. It takes work, but it is a nice way to keep your most important relations informed.

Meeting face to face is nevertheless the most effective way to communicate. It is much easier to gauge what the other thinks in a personal conversation than by exchanging emails, and meeting in person, even if only short, is always the preferred way of establishing and maintaining good relations. However, in professional contacts, it is important that you are always prepared to say the right things and to convey the right message. Sometimes, there is only limited time to do so, and hence, it is good to be prepared.

### 4.6.2 The elevator pitch: half a minute to get your message across

Imagine the following situations: By chance, you run into that important scientist whom you always would have liked to meet, and you have no more than a minute before he must leave to get his taxi, what do you say? Or a radio reporter steps up to you at a conference, sticks a microphone under your nose and asks your opinion on something? In fact, we discussed a similar situation already in Chapter 3, where we urged poster presenters to always have a 30-second summary prepared for visitors. It is for this type of occasion that you should always have a clear, succinct statement ready about what you want to communicate. It is often called the "elevator pitch," short enough to give a clear message to someone you meet by chance in the elevator, during a ride of less than a minute (although people usually don't talk in elevators).

You can think of an elevator pitch as a mini speech for raising interest in what you and your research group do, to probe interest in a new idea, a project, or a service or strength that you can offer. Doing it effectively takes preparation and experience; it is risky to rely solely on your talent for improvisation. Here are some recommendations for putting together an effective pitch.

a) Define a clear objective – what do you want to achieve? A new collaboration? Finding a talented candidate for a vacant position? A joint application for funding? An invitation to visit and give a seminar at your colleague's institution? Access to a laboratory facility?
b) Include a brief explanation of what you and your group do, the subject you work on, and the question you try to answer or the problem to solve. Make sure it reflects your enthusiasm for the subject well. Stress your strong points (your "unique selling point") and why you and your group are an attractive party to team up with.
c) State your goal briefly and clearly and leave no doubt that you have a clear plan of action – remember you are not merely asking for help, you have a proposition that can be of mutual interest (see Chapter 2; self-leaders have a plan).
d) If the pitch is used in a one-to-one contact, ensure you engage the other in some way, for example, by asking his/her opinion. If the pitch is for an audience, engage them with an open question or an invitation and make clear how they can reach you if they want to follow up. In any case, try to avoid that the interaction is

only a one-way transmission. Do not forget to present a business card or perhaps send an email to reconfirm your interest.

e) Exercise your pitch a few times, maybe record it on video, and time it. The great advantage of having gone through this pain is that you will be much better prepared to give a concise message when the opportunity arises.

While it is good practice to always have a "message in a nutshell" ready in your mind for establishing a contact, it takes a step-by-step process to turn it into a sustainable, mutually beneficial relationship. We sometimes compare the process to a tugboat that will tow a super tanker. It takes a strong and heavy cable to do this, and it is impossible to get the cable over in one go. Instead, contact is made with a much thinner rope, which is then used to pull a stronger one in, which, on its turn, is used to get the ultimate cable in that is used for towing. Building a lasting relationship works in a similar way, mutual trust and understanding develop gradually in steps. If the tugboat start pulling full force on the thin rope used for making the connection to the super tanker, it breaks. Hence, developing mutually beneficial relationships takes time.

In Chapter 6, we discuss interactions and relationships between people in much more detail and in broader perspective.

## 4.7 Conflict management

Conflicts may arise in each situation where more than one person is involved. Be it caused by a misunderstanding, a fundamental difference of views, or by incompatibility of personalities, conflicts are simply a fact of life, and they should be managed judiciously to avoid escalation and lasting damage to the functioning of your team, your organization, or even your entire department or institute, or career.

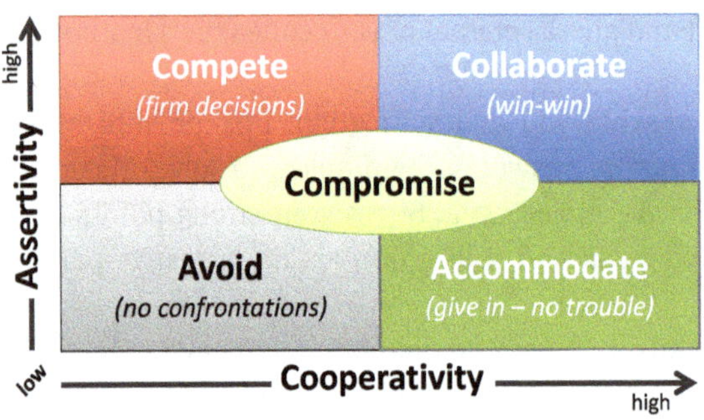


Fig. 4.6: Different ways of dealing with conflicts; adapting the approach to the specific situation is the recommended course of action.

Figure 4.6 summarizes the basic strategies for handling a conflict, between you and someone else or in your role as judge or moderator between members of your team.

It is wise to carefully adapt your strategy for resolving the conflict to the situation. For example, in case of a threatening emergency, one should be forceful and executive and take a firm decision fast. You don't want to lose time in discussions for the situation calls for a clear and immediate resolution. However, as this approach yields winners and losers, it is not wise to resolve every conflict in this way. Hence, it is only recommended for crisis situations that require immediate action.

Collaborating toward an outcome of mutual benefit, is harder to realize, but in the end much more rewarding. It implies that one should listen to either side; explore opinions, feelings, emotions, and ideas of both sides; and mold these into a resolution that is a win-win situation for all. One should probably reserve this approach of handling conflicts for truly relevant issues of strategic importance, but not for minor things.

To accommodate as a conflict strategy means you give in to avoid trouble, or you persuade one of the parties to do so. This solution is applicable in case of minor issues or if the losing party can be compensated by doing them a favor on an unrelated issue. The avoidance strategy tries to ignore conflicts or to gain time in the hope that the problem will solve itself. The purpose is to avoid confrontation at all cost. This is sometimes possible when the source of the conflict can be removed, e.g., by appointing a new employee or changing directions in an ongoing process.

Finally, one can negotiate a deal between the parties in the conflict. If both sides are willing to move a bit, it may be possible to reach a compromise that both sides can live with.

Our recommendation is that you reflect a little on the strategies for dealing with conflicts and that you try to avoid applying one approach in all cases. It is interesting to make the connection between conflict handling strategies in Fig. 4.6 and the personal operational dimensions in Fig. 2.1 in Chapter 2. The Drive dimension obviously relates to competing and using personal power, while the Feel and Do dimensions are probably more connected to less assertive strategies of avoidance and accommodation. The Doer might want to avoid and just go on; the Feeler might prefer to accommodate. The thinker might prefer the compromise approach. A combination of the Driver and the Feeler might prefer the win-win approach.

## References

Davenport, T.H.: Saving IT's soul: human centered information management. Harv Bus Rev; 1994; 72; 119–131.

Duhon, B.: It's all-in our heads. Inform; 1998; 12; 8–13.

Hindle, T.: Manage Your Time; DK Publishing Inc., New York, 1998.

Koenig, M.E.D.: What is KM? Knowledge Management Explained, 2012. http://www.kmworld.com

Manifesto for Agile Software Development, 2001, www.agilealliance.org.

Oncken Jr., W., Wass, D.L.; Management time: who's got the monkey. In: On Managing Yourself; Harv Bus Press; 1974; 52(6); 75–80.

Vohs, K.D., Redden, J.P., Rahinel, R.: Physical order produces healthy choices, generosity and conventionality, whereas disorder produces creativity. Psychol Sci; 2013; 24; 1860–1867.

# 5 Planning the road to your future in science: strategy and comprehensive plans

## 5.1 Introduction

To realize your ambitions, you wish and hope that your research group, your laboratory, and your institute operate like a team that wholeheartedly supports your vision and that things automatically go in the right direction. That takes optimal decisions about everything they have to do and operates in a way that they enhance each other's performance. This would be the ideal situation, and if you succeed in getting anywhere near, you can consider yourself extremely lucky. What would be needed to form such a self-organizing, synergetic team under your leadership?

Effective leadership involves your coworkers in a way that they work for their own goals as well as your goals. For this to happen, it must be crystal clear to all team members what these common goals are and why these are important. These goals derive from, or at least fit in, a common vision, a mission, and a strategy for what the team wants to achieve as a whole and a vision, mission, and strategy for each team member separately. In general, your own research group will have to fit in an organization, e.g., the university department that you belong to, and hence, it is important that you know what your superiors have in mind.

## 5.2 Vision

First of all, you need a grand vision that you and your most important coworkers passionately believe in, perhaps not immediately, but in any case after some time – a vision that is hopefully compelling to all people in your team and shared at all levels, a vision also that complies with the view of the larger organization you belong to (e.g., your university, or a multinational company, or one of its divisions). The vision describes the way you see your organization in context, your research group, or yourself after you have reached your goals; it is the optimal desired state in the future, say 10–25 years away from now. You could also call it a dream, but it should be a realistic one. A vision provides the inspiration for all activities and leaves room for creative interpretation and new ideas. This vision has a considerably larger span than what you want to achieve with your own organization in the coming few years.

**Example of a vision**

For example, suppose you are working in the sustainable energy field. You would then probably have a vision of what your country or your part of the world would need to do to reach a sustainable energy system in the next few decades. Your vision might be that all fossil-fired power plants are replaced by solar and wind energy parks in the next 10 years and that managing such a

https://doi.org/10.1515/9783110468892-005

transition would need significant technological development in material improvements for cheap and reliable solar cells, mechanical improvement of windmills, development of batteries for electric vehicles of all sorts, electricity grid management systems for optimal distribution of electricity, and storage systems for excess electricity on a super sunny day. In addition, the world would need trained manpower at many levels and in many disciplines. It would take decades and many research and development groups and organizations, plus industry and government, to get this all done. You can of course make only a small – although meaningful – contribution; your vision may describe this. Your research group would concentrate on one important aspect, let us say the development of solar cells, and you have clear ideas on what the technical bottlenecks are and how the associated problems could be solved, based on of a novel combination of materials. This could be your vision.

It is important to realize that your research group is part of a university department or that the laboratory you are leading may be part of a multinational company. The larger organization to which you belong will also have a vision, and it is important that you know it very well and that you have considered how your own vision relates to it. On the other hand, your students, coworkers, and staff members may also have a view on how they see their role and their future. Ideally, these visions, although probably quite different in scope, should match and not be contradictory.

The scheme below lists several questions that will help you to formulate a vision for your research group (your unit) in relation to the organization it belongs to, as well as for you personally. Your team members may use the right part of the scheme to develop a personal vision for themselves once you can present them your views for the research group.

| The Vision | | |
|---|---|---|
| **The organisation** | **Your unit** | **Yourself** |
| How does the organisation see its role in society, the world, the sector it is active<br>– What is the spirit that drives the organization<br>– In which 'market' or 'sector' will the organization operate?<br>– Which role does the organization want to fulfill and how does it want to distinguish itself?<br>– What are the organization's prime motives?<br>– Which culture for working and living would the organization like to have?<br>– What energizes the organization? | How does your unit fit in the organization and the sector in which the organization is active?<br>– What is the spirit that drives your unit?<br>– In which field will the unit be active?<br>– Which role would the unit like to fulfill and how can it distinguish itself?<br>– What are the unit's prime motives?<br>– Which culture for working and living would the unit like to have?<br>– What energizes the unit? | What is or should be your place in the organization/unit where you will be employed<br>– What is the spirit that drives you<br>– What are your prime motives?<br>– How would you like to live and work?<br>– In which field would you like to work?<br>– Which role would you like to have?<br>– How can you make a difference in your expertise<br>– What energizes you? |

## 5.3 Mission

Where your vision statement explains how you ideally see the role of your organization in a greater context, your mission statement describes what you want to achieve in the coming years. In this sense it is more down to earth and tangible than the more idealistic grand picture. The mission describes the present purpose of your organization; it specifies what it does, for whom and how, and with a focus on a shorter term of 3–5 years from now. The following scheme may help you on the way to formulate an effective mission statement.

**Example of a mission**

Continuing on the vision for research on solar cells in the sustainable energy field as described above, the mission could, for example, be that you wish to explore the applicability of a new complex material that you anticipate to have excellent properties for converting solar light into electricity and to optimize its materials properties for this purpose. After demonstrating the suitability of the material and further improving its properties, you may want to build a demonstration cell to prove that the principle works in practice and that application on a larger scale would be viable. In parallel, you may want to work on a preparation route that enables scaling up and establish contacts with industry to exploit the ideas further. You estimate that realization of this mission should be possible in a period of 5–7 years.

Note the difference in nature between the idealistic vision and the more practical mission statement. Our own experience in academia is that many prospective students looking around for a research group where they want to graduate are often more attracted by a clear picture of what the professor and his/her research group intend to do in the near term than for the bigger picture. This is logical because their mindset is generally to finish in time and with a successful thesis. Hence, having both vision and mission statements in order is essential.

| The Mission | | |
|---|---|---|
| **The organisation** | **Your unit** | **Yourself** |
| – What does the organization want to contribute<br>– What does the organization want to accomplish?<br>– Does the organization have a clear mission? | – What can your unit contribute<br>– What would the unit like to accomplish?<br>– Does the unit have a mission for itself? | – What can you contribute to the organization/unit where you will be employed<br>– What dou you want to accomplish?<br>– Do you have a mission for yourself? |

Vision and mission together contain a clear description of what you want to achieve. It may be that your coworkers or students are more interested in your mission statement or even in some concrete goals that derive from it, as it has immediate impact on their activities now. However, the organization you belong to, including your head of department or even the executive level of your university or research institute, will have a keen interest in both and especially where you want to go in the long run.

## 5.4 Goals for short-, medium-, and long-term

The next step is to derive some concrete goals from your vision and mission, for the short-term (say 1–2 years), the medium-term of a few years (in research, typically the time needed to finish a PhD project), and the longer-term, say 5–10 years from now, see Fig. 5.1. Don't be afraid to formulate these; they are not cast in stone, and you can adjust or even radically change some of the goals where necessary.

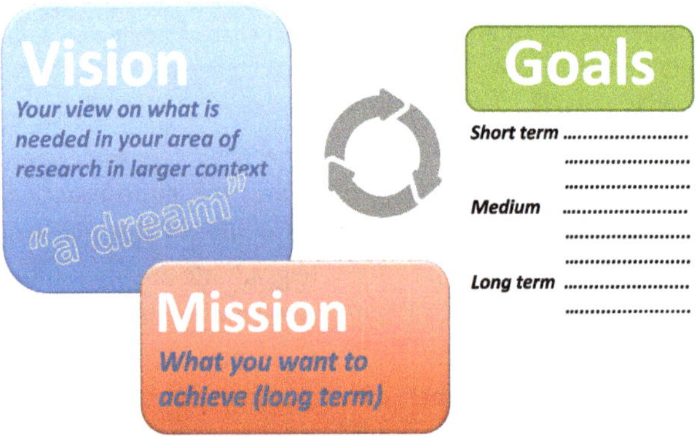


Fig. 5.1: Vision, mission, and goals for the short, medium and long term need continuous revision, and are dynamic.

### 5.4.1 Alternative views: What comes first – vision, mission, or goals?

The way we describe vision, mission, and goals above may suggest that formulating them is a linear process in time. This does not have to be the case. Many people may start from a mission of what they want to achieve or even a rather detailed set of goals, while a vision comes later. Different views on strategy development exist as well. In the Anglo-Saxon World, one often starts the process from a set of goals, while the vision is seen as a view on how to achieve mission and goals. In the Continental European setting, the vision is often more idealistic, a dream, from which mission and goals derive. It is realistic to view the set of vision, mission, and goals as dynamic, which need adjustment and refocusing along the way.

Fig. 5.2: Over time, too idealistic dreams reduce to a feasible vision, mission, and goals, in line with one's span of control.

Whatever their exact definition and role, in practice, vision, mission, and goals are never cast in concrete, but dynamic. Very often, idealistic dreams in the vision or too optimistically formulated goals need to be adjusted when reality appears harsher than anticipated or your span of control is simply not big enough (Fig. 5.2). Constraints from real life, together with progressing insight, often force us to review if the whole picture behind your plans is feasible. Later on, we discuss the concept of dynamic strategy aimed at dealing with the challenges of a rapidly changing world.

## 5.5 Strategy

For implementing the vision, we need an integral plan of action that takes all important aspects into account – a strategy for achieving your goals.

The concept of "strategy" may have different meanings and can be defined in several ways. Mintzberg distinguishes between several types, of which we mention three here (Mintzberg et al. 1998):

- Strategy as a plan – a directed course of action to achieve an intended set of goals
- Strategy as a pattern – a consistent pattern of past behavior, with a strategy realized over time rather than planned or intended. Where the realized pattern was different from the intent, Mintzberg refers to the strategy as emergent
- Strategy as a position – locating brands, products, or companies within the market, based on the conceptual framework of consumers or other stakeholders; a strategy determined primarily by factors outside the firm

In scientific research and development, one can certainly recognize these three variations, in particular the concept of emergent strategy will look familiar to many scientists. For example, one can make a beautiful research plan to investigate a certain burning scientific question. However, if one cannot obtain the funding to carry it out in its original form, one has to redefine the plan in a somewhat different direction, to be able to benefit from another funding scheme. The increasing scarcity of funds and diminishing success rates of grant proposals have urged many researchers to adapt strategies in an agile manner – a typical case of working with dynamic and emergent strategies. *Intended* strategies adapted on the basis of *emerging* strategy make *strategies dynamic* and not cast in stone. We will return to emergent and dynamic strategies later (Chapter 7, Fig. 7.5). For the moment, we limit ourselves to the concept of strategy as a plan, with building up a new research group or perhaps a new institute as the main example in mind. So for us, strategy is a plan of action, a translation of vision and mission in a plan to achieve your goals. Before we formulate it, we need to consider a few important points.

### 5.5.1 The playing field: explicit and implicit rules of the organization

In any type of sport, one needs to know the rules of the game before one can win a match. The game will have explicit rules, written out in a book, as well as implicit or unwritten rules, relating to culture, habits, or spontaneously grown customs (for example, in soccer, to kick the ball out of the field if someone from the opposite side gets hurt; it is not in the rules but every team does it). Also in science, one should be well aware of the playing fields, being the set of conditions, constraints, opportunities, and rules of the organizations in which we work or deal with, under which we will try to realize our mission. An important part of this is that we understand very well how success is defined for yourself and for the organization (e.g., your department, university, and company) to which you belong (see Chapter 1). Every organization has its own collection of explicit and implicit rules, some you may be able to challenge or even ignore, but most, you will have to obey and live with. Being aware of these can save you some frustration.

### 5.5.2 Strengths, weaknesses, opportunities, and threats: the SWOT analysis

It is essential to make a realistic assessment of your position, i.e., where your strengths and weaknesses lie, and how these combine with the opportunities or threats in your area or environment. This is the famous SWOT analysis, which is a generally accepted assessment tool in the world. The big advantage of the tool is that is does not rely on any theory and represents an objective, holistic method to evaluate one's position or one's chances. It is applicable to an individual, a group of people, an organization, or even an entire country. Figure. 5.3 shows a SWOT diagram. The fields Strength and Weaknesses are internal factors and can, for example, be affected by changing the composition of a team or finding strong collaborators, while Opportunities and Threats are mostly external factors, outside one's span of control, dictated by the outside world. The appendix has a worksheet that can be used as a template for your own SWOT analysis.

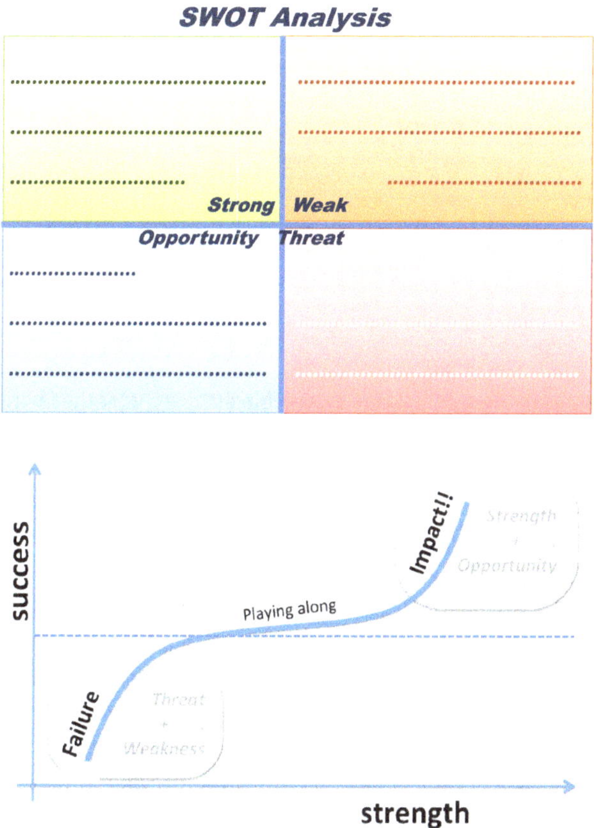


Fig. 5.3: The SWOT analysis is a generally accepted tool to assess your position and chances. Below: Combining your strengths with favorable opportunities is your best bet while you wish to avoid areas where threats might harm you extra on your less strong points.

**Failed proposal due to violating the SWOT analysis**

It sounds almost silly in Fig. 5.4 to assume that one would venture into an opportunity without having particular strengths, but it happens often in science. Prof. John tried several times to obtain grants outside his particular field of expertise during the course of his career. In 2010, funding opportunities were dwindling in his country, until a fairly substantial funding scheme for the chemistry of converting natural sources such as lignin and cellulose from plants into base chemicals and fuels. As a materials chemist, more versed in solid-state aspects of matter than in organic reactions, he submitted a proposal on developing new catalysts. It was rated "good and worthy of granting when funds are sufficient" but not as "excellent" or even "very good," which was the threshold for granting due to oversubscription of the program. This is a typical example of going against the principles of the SWOT analysis. Had he teamed up with a specialist in organic reactions, he might have had a better chance.

| | Opportunity | Threat |
|---|---|---|
| **Strength** | *Consider to go for it* | *Avoid* |
| **Weakness** | *Don't do it* | *Defend yourself* |

Fig. 5.4: How to base an action plan on the outcome of a SWOT analysis.

### 5.5.3 A comprehensive plan

If your vision/mission/goals, your thoughts about a possible strategy, together with the insight in how external opportunities and threats in combination with your strengths and weaknesses, have led to a more fixed idea about your plans, it is time to formalize all this in a comprehensive plan, as given in Fig. 5.5. This scheme, entitled "My Organization: Why – How – What" can be a tremendously valuable tool in planning and also in analyzing where you stand at any given moment.

The scheme is built on two axes: vertically, it runs from the philosophy behind your strategy at the top to the practical realization at the bottom, while horizontally, it ranges from internal matters such as vision, culture, and projects to your external visibility of your organization in the public relations and the output you produce and will eventually give you your recognition from the outside world. On the bottom of the scheme are the enablers for your work, namely, people, infrastructure and funds. We will go through each of the headings separately:

- **Vision**: a short version of your vision or mission, or even your goals (as you prefer). This part has its influence on all other sections of the scheme.
- **Culture**: the values of your organization, the way of thinking, behaving, and working together as you would like to see it in your research group, your department, or your organization; also think about core values such as safety, ethics and integrity, open exchange of ideas, mutual respect, etc.
- **Projects**: the way your vision on, e.g., research and education is translated into practice, the collection of activities of your organization such as specific research projects, and the courses you teach, perhaps also initiatives to structure your discipline on a national or even larger scale.
- **Strategy**: a brief summary of how you want to realize your vision/mission/goals.
- **Management**: the way you structure all the necessary activities in your unit, such as teaching and supervision of students, who is/are responsible for safety, who oversees the state of the infrastructure and supplies, how funding and reporting is organized, whether there is a management team with regular meetings, etc.

Fig. 5.5: A comprehensive plan addresses all important aspects of your organization.

- **Organization**: this is a short description of important procedures for your group (a "mini Soldier's Handbook"), with, for example, how you arrange project and progress meetings, periodic seminars, management team meetings, procedures for safety checks of your laboratories, budgeting, reporting, monitoring literature, etc. Keep it simple and probably as lean as possible, but certain aspects need to be arranged properly.
- "**Market Strategy**": this is the philosophy behind your strategy to let the world know what you are doing and how successful you are and how you intend to use contact and relations management (Chapter 4). You identify target audiences (stakeholders such as industries, companies, funding agencies, professional societies, your peers in your discipline, the students you would like to reach and attract to your group) and why and how you want to inform them with what sort of frequency. Please realize that just publishing scientific papers or giving the occasional talk at a conference may not be enough to distinguish yourself (see also Chapter 3 on communication and presenting science in this book).
- **Public Relations**: how you realize your "Market Strategy", i.e.", a scheme for informing different stakeholders of your successes. Examples are an attractive and informative website that is regularly updated, regular visits to companies, twice-per-year popular article for the larger public (to appear in a local or national newspaper or a magazine of the scientific community), and a press communiqué to publicize a truly remarkable finding or the fact that you managed to publish a paper accepted in a highly ranked journal. Also think about a policy for keeping students informed

about your activities, as you hope that the best ones will choose your group for their final projects or your department to do their master's or PhD. Being present on the social media or having short video highlights of your group's activities can also be very effective. Finally, having a plan for when and where you or your coworkers and students present at conferences is also essential for promoting your work in the community, to obtain feedback and new ideas and to make new contacts.
- **Output and Results**: here, you probably think first of the scientific papers and reports that you publish, the patent applications that you file, the designs you present at an exhibition, the posters and presentations by you or your students at conferences, and the abstracts in the program book. However, also the students who graduate from your group or your department or the postdocs who leave for academic or industrial positions count as output of your activities (and have probably much more value than publications in a scientific journal!).
- **People, Infrastructure, Funding**: At the bottom of the scheme are your intended resources: where and how you intend to recruit your staff and attract your students, how you want to build up your laboratory infra structure, and finally where the funding may (or will) come from.

Looking at the upper part of the comprehensive plan in Fig. 5.5, we see that it has three columns: the left one for the internal aspects (vision/mission, culture, and projects), the middle for operations (strategy, management structure, and daily organization), and the right column for your external visibility. It is the latter that will lead to recognition for your endeavors, and therefore, it is of great importance for your chances to obtain funding for excellent infrastructure and top-quality people who like to work with you and also for giving you a good position for valuable collaborations that will widen your opportunities to be successful in science.

### 5.5.4 A procedure for assessing your research unit, your department, and your organization together with your coworkers

The scheme of Fig. 5.5 can well be used to reflect on your present activities, either by yourself or, even better, together with some of your coworkers. We describe an effective procedure that both authors have used many times in assessments of all kind of organizations, ranging from a small research group to large governmental research institutes and even large companies.

1) Invite some of your team members to a half-day meeting, preferably "off-site" and with mobile phones switched off, to minimize distractions. It works best if you involve members from different levels, e.g., for a small research group, perhaps everybody, and for a typical European academic research group, the permanent scientific staff, a technician, the secretary of administrative assistant, and definitely one or more students.

Fig. 5.6: Each note mentions one issue, with colors green, yellow, and red indicating success, needing attention, or failure, respectively. In a real situation, one typically has 50–100 of such notes.

2) You will need a pile of sticky notes in three colors, green, yellow, and red, for example, and enough felt pens and a poster-sized version of the "My Organization" scheme of Fig. 5.5. Ask the participants to take 15 minutes to think about the organization and its activities and write down what issues, problems, and successes come to their minds: One subject per sticky note, green if it counts as successful, yellow if it needs attention, and red if it goes wrong. Anything that comes to mind is good, on every level, be it a daily problem in the laboratory, an administrative hick-up, a successful appearance in a conference, the failure of a grant proposal, student complaints about a difficult exam, a perceived shortcoming in the vision behind your organization, everything that comes up is important, no matter at what scale or level. Figure 5.6 gives an idea.
3) When all have finished writing, the notes are collected, briefly discussed one by one, and each note is placed on a poster version of Fig. 5.5, in the category where it belongs. People cannot withdraw notes, but they may add new ones, stimulated by the discussion. It is important that you as a leader of the discussion does not respond too much or start to defend yourself. The aim is to get all issues on the table that your team members find important. Whether negative or positive, fair or not, the point is deemed relevant enough to be brought up by your team members, so you have to appreciate it as important. Also see it as a comment on the organization, not on you personally.

4) At the end of the discussion, you will have the "My Organization Poster" clotted with colored notes, perhaps looking as the example in Fig. 5.7. What does it tell you? In the first, place where the team's focus is and if you are paying sufficient attention to all the important aspects of running your organization. Taking Fig. 5.7 as an example, it appears to reflect a team of doers and drivers, with a strong focus on projects and output. The green notes indicate that most of projects are going well, and the predominance of green on the output side give evidence that the group can boast some successes. The culture is evidently pleasant and very likely inspiring as the positive feelings on the team's output indicate. However, the daily organization and the management structures seem to lack structure; perhaps, the group leader relies too much on improvisation or allows urgency as the main driver here. Vision and strategy clearly need more attention, as the team does not seem very aware of it. Of course, it is always possible that the leader has the vision, mission, and strategy clearly in his/her mind for himself/herself, but sharing it more with the members may be good and a first step toward getting the daily organization in better shape. Also note that a strategy for public relations is mostly missing and that one or two persons raised concerns whether the team receives adequate recognition for its endeavors. The resources clearly need attention, with acquisition of funding as the most important concern.

Fig. 5.7: The team meeting leads to an overview of all issues shared by the team. The distribution indicates clearly where the focus of the group is, and which aspects need more attention.

## 5.6 Concluding remarks on vision, mission, and strategy

To summarize this chapter, as an accomplished self-leader and as the leader of an organization, it is essential that you have a compelling and well-articulated vision and mission. While the vision reflects your dreams and goals in the longer terms, your coworkers (students) may be more interested in your mission, which is in essence a more concrete set of plans and objectives for the coming years. The strategy is a plan of action on how to accomplish your goals, with a comprehensive view on internal aspects (vision, culture, and projects), daily operations (tactics, management, and organization), and how to present your achievements to the world. The latter is of crucial importance for your chances to obtain the necessary resources (funds, infrastructure, and people) for realizing your dreams.

## Reference

Mintzberg, H., Ahlstrand, B., Lampel, J.: Strategy Safari: A Guided Tour through the Wilds of Strategic Management; The Free Press, New York, 1998.

# 6 Understanding how to build successful teams: creating synergy between people who trust each other

## 6.1 Modern science is teamwork

Modern research and development is teamwork. Unlike a century ago, one can hardly do anything alone. Professor Hutchings at Cardiff University runs a hugely successful academic research institute in catalysis with many connections to other laboratories in the world. Often, 10–15 people coauthor his papers, and sometimes more than half of them are from elsewhere (see for example Hughes et al. 2005). Another good example is our colleague Professor Jens Nørskov at Stanford University, a famous theoretician, who also has a track record of mostly multiauthored papers, with coauthors from different places and disciplines (see for example Nørskov et al. 2002). In our own research, although positioned at the fundamental end of the spectrum, we have always very fruitfully collaborated with large chemical companies such as Shell, Sasol, and Synfuels China Technology. Industrial researchers know from experience where the real problems and questions are, but they cannot usually afford to delve into the fundamental aspects underneath, and that is where academic research groups have their strengths. On the other hand, our industrial colleagues are also very good at telling us if our efforts in research make sense or that we trying to find solutions for irrelevant problems. Or as our distinguished mentor Dr. Jens Rostrup-Nielsen often warned us (Rostrup-Nielsen 2016): “University professors should beware of non-applicable applied research.”

But also on the smaller scale of the academic research group, consisting of a professor with one or two postdocs and a few PhD students, or an industrial project group, the benefits of effective collaboration and teamwork are indispensable.

What makes a team successful? In any case, the members should share an interest in the same goals and believe in similar values, among which trust, honesty, modesty, and discipline are probably the most important ones. Finally, the members bring talents, expertise, and know-how that are mostly complementary, with perhaps some overlap to enable mutual understanding. Finally, effective leadership plays an important role, although in self-propelling teams, the members lead intermittently, as the situation demands.

According to Lencioni (Lencioni 2002), members of a cohesive team exhibiting positive synergy behave as follows:

1. They trust each other.
2. They openly discuss ideas critically in a constructive way.
3. They commit to decisions and plans of action.
4. They hold one another accountable for delivering against those plans.
5. They focus on achieving collective results.

https://doi.org/10.1515/9783110468892-006

Points 2 to 5 of Lencioni regard tangible aspects that are very important for producing concrete results. Point 1 refers to the intangible aspect of trust. People have inside them a lot of hidden potential, in the form of talents and thrives, as well as willingness to connect and share. When aware of this, they can lead themselves from within with dedication and inspiration to make their specific contributions and a productive process of co-creation and inter-creation may start. If a leader succeeds in combining the intangible value of human sources with the tangible process of creating results in a team, it can reach high levels of performance and impact of the work.

Therefore, a team, whether it consists of your students and postdocs or your experienced colleagues, will have better chances of success when the members feel that they are fully accepted as they are, and thus really belong to the group, and that they are valued for their contributions. This is the so-called 3B Principle: **b**e oneself, **b**elong to the group, and **b**e valued, as we discuss later in this book.

### 6.1.1 Hierarchy in teams

Self-propelling teams change leaders all the time, sometimes many times per day, and perhaps even without noticing it. Oshry (Oshry 2007) presented an interesting system model that illustrates the principle well (see Fig. 6.1). The team members have many roles, which can be classified as top, middle, bottom, or customer, which can be seen as conditions that all members face, irrespective of their formal position or function in the organization. We are "top" when we are accountable (i.e., possess assigned responsibility) for assignments; "bottom" when we fully depend on others and all we can do is help them (we are fixers); "middle" when we are intermediate between at least two others and, for example, resolve their conflicting needs, demands, and priorities; and "customers" when we need the output of the activities of the team or some of its members in order to move ahead. In a sense, as a customer, we serve as the validator of the activities, while as intermediates in the middle position, we can play an integrating role. In the team, we play all these roles interchangeably and simultaneously, depending on what the circumstances call for. It is important to realize that in each of these conditions, we, as accomplished self-leaders, make essential contributions to the success of the team.

Fig. 6.1: In Oshry's system model for teams, the members continuously move in and out conditions such as top, middle, bottom, and customer, irrespective of their formal function or position in the organization.

**Sports teams**

*Team sports such as basketball, volleyball, and soccer essentially rely on the talent, skills and self-leadership of the players. Although the coach is of great importance for tactics, strategy, deciding who plays, and all that, during the match, the team has to take all instant decisions and to rely on an interesting combination of experience, routine patterns, genial hunches, and hard work to win a game. The self-*propelling *team relies on automatisms, occasional brilliant actions, and the determination to win whatever happens. The team has to win the match, while the coach stands on the sideline.*

## 6.2 Understanding the people you work with – we are all different

Having the wrong person on board can interrupt or, worse, paralyze or even break up an entire team. How do you know beforehand if a candidate (e.g., an applicant for a new position in your group) will fit in? First impressions count strongly, especially when a new student or an applicant comes to see you about a position in your research group. We all will have an immediate impression of somebody we meet for the first time, and whether it is right or wrong, incomplete, or affected by the occasion (think, for example, about the possible stress of a candidate for a formal job interview or the talented actor who plays a role), it determines our opinion about this person for some time, until we get to him/her better. Experienced recruiters are very well aware of this and try to base hiring decisions on more objective criteria. Of course, the decision to go ahead with the student/applicant is a very important one. Apart from the candidate's knowledge, experience, skills, and expertise, it is extremely important to judge if he/she will fit in your team and can interact positively with you and the other members of your group.

### 6.2.1 A little psychology: the "big five" personality traits

Without any claim of treating this subject correctly in any depth, we believe that a little insight in psychology of the human character is helpful. While the classification Feel – Think – Do – Drive that we introduced in Chapter 2 has shown its value in classifying people's behavior in the work place, psychology deals more broadly with people's mental and behavioral characteristics. Historically, the ancient physician Hippocrates (460–370 BC) formulated a theory based on the notion that having too much or too little of four essential body fluids determines one's personality. It has later been associated with the classic elements earth, water, air, and fire. Although largely abandoned, one can still recognize its effect in older literature, opera, and drama, and it is interesting to know at least its origin, see a short summary in Tab. 6.1 (Childs 2009; Eysenck 1967; Steiner 2008). Interestingly, the typification bears some similarity with the operational personality and management dimensions: Air – Thinker – Content, Fire – Power of will – Results, Earth – Doer – Operations, Water – Feeler – Relations.

Tab. 6.1: Hippocrates' theory of body fluids determining one's personality.

| Type | Body fluid | Element | Characteristics | Comments |
|---|---|---|---|---|
| Sanguinistic | Blood | Air | Optimistic, active, social, creative | Notorious late comers; can be forgetful |
| Choleric | Yellow bile | Fire | Short tempered, fast, extrovert, egocentric, task oriented | Often have management potential |
| Melancholic | Black bile | Earth | Analytical, wise, introvert, conscientious, individualistic | Down-to-earth, objective attitude, often susceptible to negativism |
| Phlegmatic | Phlegm | Water | Calm, patient, thoughtful, tolerant | Innerly oriented and motivated |

Much more widely accepted nowadays than the obsolete body-fluids theory is the so-called Big Five or OCEAN model, which describes personalities not so much in terms of characteristics rather than in five traits (the first letters form the acronym OCEAN; Fig. 6.2):

- **Openness** to new experiences and ideas, versus preference for the comfortable and safe well known. Personalities scoring high on this trait are usually interested in modern art, literature, and traveling to new destinations and curious about new insights, latest fashion, and gadgets, while persons on the other end of the spectrum prefer to stick to their trusted routines and are not so keen on change. Of course, we hope to have scientific researchers in the first category, but please do not underestimate the usefulness of team members who value and safeguard well-established routines for maintaining quality and safety standards and sound research practices in your laboratory.
- **Conscientiousness**, meaning acting according to careful planning and being accurate without forgetting details, versus improvisation and spontaneous action, with the risk that important aspects are overlooked. As a leader of a research group, we obviously wish to rely on our coworkers to do their research accurately and completely, but it will not be the first time that following a spontaneous idea for a crazy experiment eventually leads to unexpected and extremely valuable results and insights.
- **Extraversion,** being interaction and people oriented, versus being introvert and internally oriented. Extravert personalities are often also perceived as assertive, energetic, and full of initiative. However, we also know the stereotype of the eccentric scientist, who prefers to work alone, reads and reflects a lot, and can be very successful.

- **Agreeableness,** being friendly, compassionate with others, and seeking compromises, versus competitive, self-oriented, and inclined to distrust others until the opposite is proven. Obviously, a team will be happy with some "agreeable" types, who make all others feel welcome and at their place. However, if the whole team is all the time "agreeable" and avoiding conflict, no one will ever bite the bullet and difficult issues may remain unresolved.
- **Neuroticism, or emotional instability,** implying sensitivity to stress, feelings of insecurity, and tendency to amplify negative stimuli, versus being relaxed, balanced, and realistic. High emotional stability reflects itself in rather calm personalities, who on the other hand might be perceived as unconcerned or uninspired, while on the other end of the spectrum, we find personalities that are easily excited and seen as dynamic and perhaps a bit unpredictable.

The principle that the five OCEAN traits form a basis set to describe human character has been widely accepted, even across different cultures, although regional adaptations do exist. We think it is important to stress the value of having team members with a healthy variation in their personality traits.

For scientific researchers and in particular scientific leaders, we can envisage a few other traits that are important as well, for example, intelligence, creativity, endurance, and not to forget, integrity. A theory called "trait leadership" (Zaccaro 2007) identifies no less than 17 assets of successful leaders, including the OCEAN Big Five, a number of typical management skills, as well as charisma, described as the ability to inspire their environment based on a compelling vision, that induces enthusiasm and commitment.

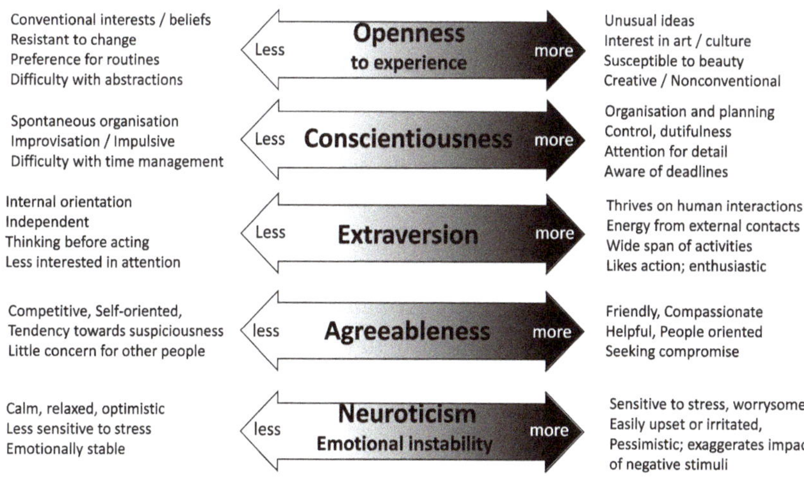


Fig. 6.2: The five traits, often called the "Big Five," that are used in psychology to describe characters. Together, they form the acronym OCEAN.

### 6.2.2 The Myers-Briggs types

The Myers-Briggs Typification system[1] distinguishes four dichotomies on which people act, take decisions, or approach situations, namely, **E**xtraversion – **I**ntroversion, **S**ensing – **IN**tuition, **T**hinking – **F**eeling, **J**udging – **P**erceiving, and then identifies which property is dominant in a person. This leads to 16 different personality types, indicated with four letters, like ESTJ, INTP, ENFJ, etc. The system has found international acceptance and has found its way to recruitment and selection procedures. The disadvantage is that it is almost a binary system, one is either a feeler or a thinker, and subtle mixtures of these characteristics are not possible in this classification, even apart from the question whether feeling and thinking should be considered opposites on the same scale.

### 6.2.3 Personal operational and self-leadership dimensions

Although not beyond all possible criticism either, we prefer to work with the simpler four personal operational dimensions that we introduced in Chapter 2, namely Feel, Think, Do and Drive. They provide a simple but practically useful system for describing and understanding the behavior of people in their work and, to some extent, their interaction with other people, as we explore in the next pages. We often refer to these four as management dimensions.

We use these four dimensions in our (self) management assessments. At one axis are the personality dimensions (doer, thinker, driver, and feeler), reflecting the thrives and preferences that people have. This is what most assessments do, and it is indeed valuable to understand how someone functions in his/her natural way. However, these assessments give no indication if people are in control of what they do. For example, whether a person is extravert or introvert does not reflect at all whether he/she is capable and successful. Our assessments therefore have another axis: the management dimensions (operations, content, results, and relations) indicate whether people are in control of how they are expected to perform daily.

Attempting to capture aspects of personality, we add three personal (self) leadership dimensions: the "*I*", the "*Self*," and the "*Impetus*"[2]:

- The "I" is externally oriented, individual, and tangible; is aware of what is going on and needed in the here and now; and is oriented directly at concrete outcomes, results or output. For this, the "I" is aware that a person needs a tangible position, with power, status, respect, and recognition, and that a person will strive for the interest of himself/herself and the organization he/she belongs to.

---

**1** The Myers & Briggs Foundation. http://www.myersbriggs.org/

**2** Impetus is defined as a force that causes a process or an activity to be done, such as impetus for improvement, impetus for further study, etc.

- The “Self” is internally oriented, intangible, and aware of what is needed in the there and then; is directed at processes; and believes in the automatic outcomes, results, and output that these will deliver. For this, the “Self” is aware that a person needs trust, belief, recognition, and a connection with the inner sources of know-how, wisdom, and intuition and that a human will strive for harmony and needs like being here, belonging to the collective, and being able to make a contribution.
- The “Impetus” (sometimes also called “Spirit”) believes in connecting to meaningful purposes and sources of positivism and wisdom, thus creating enthusiasm and inspiration for everybody. It is strongly connected to one’s thrives, ideals, and personal sources of inspiration.

#### 6.2.3.1 Interaction between people based on operational personality dimensions

Scientists in the exact domains of mathematics, physics, chemistry, life sciences, and engineering are generally not so interested to incorporate psychological considerations in their thinking and way of working. We nevertheless believe that an open mind for how personalities may differ and appreciation for what drives people help a lot in building productive collaborations. We have seen many examples of great and highly successful scientists who, over the years, have developed into real people managers, who know how to inspire their team members to excel in many ways, and often established warm and long-lasting personal relations with them.

Figure 6.3 shows the profiles of two people with entirely different operational characteristics. The question is now: Can these two work effectively together? Well, at first sight, the profiles seem to complement each other, so if the Do-Drive oriented person would be willing to carry out the ideas of the Feeler-Thinker, then this could be a good team, but not necessarily so.

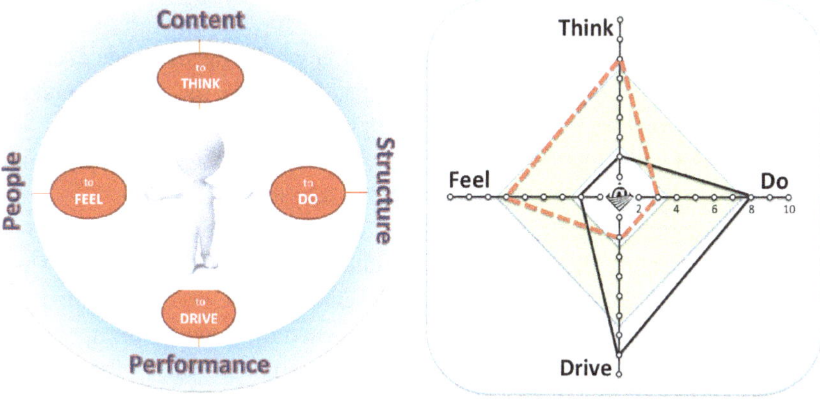


Fig. 6.3: The four operational personality dimensions along with the profiles of two clearly different persons.

Let's suppose you just had a progress meeting with your team. You ended with a list of tasks for the team in the following days and you agreed on who does what.

The DOers among your team members immediately start to carry out their tasks, e.g., ordering materials for the new project, making changes in the laboratory, and filing data from the previous project. Their working mode is to be active and do things as much as possible and they can do so because there are clear protocols or routines for their activities.

The THINKER approaches the work package by first reflecting on what is needed, how it is best done, perhaps makes a plan and then starts carrying it out. Along the day, he/she frequently stops to think: Am I doing the right thing, what are the alternatives, should I do it differently, etc.? He/she thinks first and reflects a lot on what is happening.

The FEELER acts on sensitivity and subtle consciousness, intuition, and possibly, experience. He/she feels what needs to be done, is willing to improvise, interacts a lot with other people, and gets energy out of these contacts. The FEELER may involve other people to help, and because he/she has friends, the work will be done.

The DRIVER, a typical achiever, has a strong will to get the job done and has a strong intuition for what is needed for getting results fast. He/she thrives on generating tangible results and ticking off items from the to-do and deliver list, and in this, he/she can sometimes be quite dominant toward other team members.

Successful teams have members who are driven by different dimensions. Suppose your group has only people whose first impulse is to DO things. Think of people who like to synthesize new materials, who love to measure properties, to collect literature, to clean the laboratories, etc.? What if they run off to do these things, but nobody THINKS carefully if these are the right things to do, or nobody has the intuition to FEEL that the activities are not so clever? Balanced teams will combine a few approaches, and hopefully correct each other, and together go in the right direction, whatever that is.

It is interesting to contemplate a little on what happens if two personalities with their main operational personality dimensions interact. Suppose a DOer and a THINKer have to carry out a task together. The positive (i.e., synergetic) mode is that the THINKer comes up with a clever way to execute the task, while the DOer starts by trying out a few different approaches, after which together they decide on the best way and they complete the job. In the negative mode, the THINKer makes it too complex, and the DOer loses interest to work with him/her.

A typical ineffective interaction occurs if two DOers together run off and "start doing," without any form of contemplation if what they do makes sense.

Interesting combinations are those where a DRIVER is involved. This type of personality can be quite dominant to make others work for him/her, as he/she is determined to get the job done and produce tangible results. When combined with a THINKer, effective modes of collaboration result when the latter comes up with the best and perhaps some alternative methods for completing the job, while giving the DRIVER the feeling that he/she is in control. If the outcome is positive, both benefit (and let's hope the important contributions are acknowledged). However, ineffective interaction is also conceivable, for example, if the THINKER insists on one strategy and the DRIVER is not patient enough to try it out because he/she sees a quicker way.

Another example is the following. You, a DRIVER for achievement, have a strong urge to get a difficult project finished, and your coworker is a typical FEELer. Synergy is likely to occur if the other accepts your lead and helps you, while you value the other's role and make him/her feel appreciated. The ineffective interaction is not difficult to imagine: your dominance is perceived as lack of recognition for the other's contribution and is seen as unfair. Table 6.2 summarizes the constructive and destructive modes of interaction between team members with examples. It can be a useful exercise to construct such a table for the members of your team.

Tab. 6.2: Interaction between you and people with certain operational personality dimensions, seen from your perspective. Green: effective and Red: non-productive.

| | | The other | | | |
|---|---|---|---|---|---|
| | | Do-er | Thinker | Driver | Feeler |
| You | Do-er | Agreeing what to do & doing it | Realizing the good ideas | Executing the orders | Taking care for work |
| | | No reflection on what is done | Superficiality; lack of understanding | No feedback or too late | Forgetting to establish real personal contact |
| | Thinker | Developing and proposing feasible solutions | Agreeing on content and strategy | Providing alternatives; offering sense of control | Gaining trust through knowledge and integrity |
| | | Making it too complex | Too little awareness of realistic context | Insisting to be right, on having the final word | Focus on content without good personal contact |
| | Driver | Offering support for implementation | Acknowledging insight; offering to implement it | Swift and energetic action | Acting responsably and accountably |
| | | Superficial pragmatism | Ignoring content; overruling the other | Competition; unvalidated judgements | Dominance; perceived as unfair by the other |
| | Feeler | Taking care that tasks are carried out properly | Being interested in the other's vision | Utilizing his strive for dominance | Acting on a basis of mutual understanding |
| | | Losing reality in focus on execution | Not enough feedback on content; no output | Emphasis on pleasing the other | Missing practical relevance |

Effective Interaction | Ineffective Interaction

## 6.3 Successful team leaders have emotional intelligence

How can a leader ensure that the team thrives and that the members work in an atmosphere of harmony and synergy? In Chapter 2, we briefly summarized the characteristics of successful leadership as providing a compelling vision shared with and by the entire team, having a highly developed sense of emotional intelligence and being good at personal relationships (including mutual trust and respect), and disposing over ample managerial skills. We still have to explain what emotional intelligence is and how we recognize it in people.

Goleman has given a rather rigorous description of what emotional intelligence entails (Goleman 2000). He defines the concept as a complex of five ingredients, being self-awareness, self-regulation, motivation, empathy, and social skills. We largely follow his treatment here and go through each of these qualities.

- Self-awareness: ability to recognize and understand one's own moods, emotions, drives, and their effect on others. Persons with self-awareness are usually confident, have a realistic impression of themselves, and often display a self-deprecating sense of humor.
- Self-regulation: the ability to control one's own emotions, particularly if these can have negative impact on others (anger, drift, dislike, etc.), and to suspend immediate judgments (think before acting). Self-regulators can often cope well with ambiguity and are open to change, or even to the unknown.
- Motivation: passion to realize a vision for idealistic reasons rather than for salary or status and an inner drive to accomplish goals persistently and optimistically.
- Empathy: the ability to recognize and understand the feelings and emotions of others and to interact accordingly with them. Empathic leaders are aware of cultural aspects and sensitivities and are often successful in developing and retaining talent. Servant leadership is the style where empathy is most prominently present as an essential quality.
- Social skills: being good at establishing and managing relationships and building networks, and finding common ground among team members and/or stakeholders in general. As a leader this person knows how to persuade others, to induce change, and to build and lead high performing teams.

Understanding what emotional intelligence is and the role it plays is particularly important if your team is international. Highly developed emotional intelligence can be needed to understand and guide people with different cultural backgrounds, often with their own values that are different from yours.

Deutschendorff (Deutschendorf 2013) has given a crisp summary by stating that emotionally intelligent leaders are generally non-defensive and open to their environment, they are well aware of their own emotions and know how to control these, while they readily recognize the emotional state of others and respond appropriately. Furthermore, they are available for the people reporting to them, not necessarily by always having their office doors open, but by offering clear and non-forbidding procedures for quick contact when needed. Finally, and obviously a clear hallmark for successful leadership, emotionally intelligent leaders are able to check and control their ego and allow their coworkers to shine.[3]

3 A self leadership assessment adressing emotional intelligence is available from www.scientificleaders.com.

### 6.3.1 Emotional intelligence and emotional culture are linked

Chances are high that the emotional intelligence of the leader sets the tone for the emotional culture in the group (Barsade & O'Neill 2016). Positive feelings shared by the group translate to better performance and higher quality of work, while chances of burnout and people quitting jobs prematurely are lower. When negative emotions dominate in the group, think of fear to make mistakes, anger about working conditions or lacking attention from the management, or right out suppression, the commitment of the team will be lower, while some members may even become indifferent about reaching the leader's goals. In essence, such situations lead to behavior focused on not making mistakes, hiding talents, and suppression of creativity, involvement, and dedication.

Note that the cognitive culture can easily be described in a tangible way in terms of the shared intellectual values, norms, artifacts, and assumptions on which the team operates in daily life. However, the emotional culture is mostly transmitted through less tangible channels, such as body language and facial expressions. If the boss always looks worried or angry (whether on purpose or not), the impression may have an overriding effect on the atmosphere or emotional culture of the group.

The emotionally intelligent boss will be aware of such undesired effects. He/she will also understand the great value of positive "micro moments" in daily life. A brief inquiry about someone's health after last week's flu, a compliment for something done well, a bit of small talk about last night's football match, a little gesture on somebody's birthday – it does not cost much, but it does wonders for the emotional climate. The same holds for the entire team, of course. Practicing good manners, like saying "please" and "thank you," is a simple enabler for people to be able to work together, whether they like each other or not (Drucker 2008). Having photos of the last group outing on the wall will also do better than a poster with do's and don'ts and a list of consequences when the rules are broken. Hence, well-developed emotional intelligence of the leader and the team will very likely translate to a positive and stimulating emotional culture in the workplace, where people feel confident enough to behave authentically.

## 6.4 The "3B" approach to authentic resources

In a team, mutual feelings of trust and respect, of belonging to the group, and of being valued for what one contributes are key factors for people to become successful team players. We have coined this "the 3B Approach to Free Authentic Resources" and it is based on the notion that feeling valued in relationships is often a key for successful performance and personal satisfaction. These "3Bs" are as follows:

1) **Be** one's self, meaning that people feel so confident in and accepted by their environment that they behave fully authentically. There is no need to play a role,

to hide weaknesses, to pretend that one is better than one actually is. The comfortable feeling of being accepted in a group stands opposite to the fear of being rejected, which leads to behavior focused on not making mistakes.

2) **B**elong to the group, the opposite of feeling abandoned or locked out. Feeling fully accepted by the group will undoubtedly help to use one's authentic resources as meant in the previous point. Members of the team will not have to fake their behavior and attitudes to be accepted.
3) **Be** valued, that is feeling recognized for one's contributions and feeling confident that one's strengths are appreciated. At the opposite end of the spectrum, one would, for example, have the situation that a team member develops an inferiority complex due to the feeling of missing skills, or not knowing enough.

Scientific leaders, or in fact any leader, should thus ensure that their students and coworkers experience that they are allowed to be who they are and that they form accepted members of the group. By guiding them and challenging them to contribute in small but oversee-able portions, they will learn step by step and simultaneously build on an authentic stable personality. The fundament for authenticity is recognition, which is a matter of taking people seriously for who they are and what they think and want. Listening is an important skill, just as giving attention and constructive feedback. When successful, the approach leads to an atmosphere where co-creation through natural interaction can take place. Dialogues help the other to be clear on his/her intentions, feelings, and emotions. Discussions could help the other how he/she would like to reach his/her intention.

## 6.5 Supervision – helping students to become colleagues

Graduate students (i.e., MSc and PhD students) form the core of academic research groups, thus making their proper supervision a substantial task of the academic research scientist. Many universities offer courses on supervision for their staff, and excellent literature is available (Wisker 2012).

According to Lee (Lee 2008), supervision of doctoral students encompasses the following ingredients and concepts:

- functionality – guiding students to manage their research projects;
- enculturation – introducing students into the community of their scientific discipline;
- critical thinking – stimulating students to critically question and analyze their work;
- emancipation – encouraging students to question and develop themselves; and
- development of quality relationships – where students are enthused, inspired, and cared for, also ensuring that they move along a path toward increasing independence.

For the authors, the ultimate success of supervision and guidance is if our students, new employees, or younger team members develop into colleagues whom we regard as equal on the subject of their activities and whose knowledge and/or expertise is actually superior to those of our own. How is this achieved? Several factors play an important role. Is your relation based on hierarchy, on power, or – opposed to this – on mutual trust between you and your protégé, in an atmosphere of openness and honesty and shared desire to do what is the best for the organization and for the two of you? The latter would be ideal, of course. It may take a while before a student recognizes this and realizes what your real expectations are. We all know the timid types who come with the attitude to learn mostly facts and procedures and obediently do what the professors tell them to do. If that is the dominant culture in your environment, one should not expect too many initiatives from the group. Stimulating them to be more assertive, confident, and hopefully, become true self-leaders may require much effort from you. Again, trust and feelings to belong to the group and to be appreciated for what they contribute are key factors for many students to become successful team players.

As illustrated in Fig. 6.4, adapted from the work of Blanchard and coworkers (Blanchard et al. 1985), a new student, employee, member of the team first needs direction, and instruction, as his/her competence level is low, but when he/she gradually gains experience and confidence, the supervisor can adopt a more coaching and

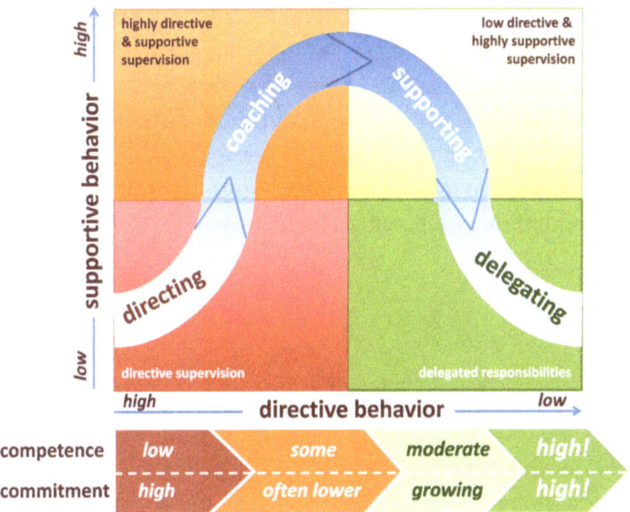


Fig. 6.4: Trajectory of becoming acquainted with tasks, for example, for a student entering the research phase of his/her studies or a new member of a project team. The novice first needs direction, but gradually the supervision style can change to coaching and supporting, while after some time, tasks can simply be delegated. All the time his/her level of competence increases, but the supervisor should be aware that commitment may vary in stages when the job presents difficulties or is simply different than expected (figure adapted from Blanchard, Zigarmi, and Zigarmi).

supportive role, until the team member feels so well in place that he/she becomes self-propelling. Note the ever-growing level of competence, while commitment and confidence may vary, for example, a few months after the start when reality strikes and the new team member discovers that the life of a researcher can be tough and different from what is expected. A few successes help to reestablish confidence and commitment. Supervisors should be aware of this often occurring situation.

When competences are low, your coworkers need you to set clear goals, to make a plan of what to do, and to set priorities. They need training to become skilled and monitoring to check progress or to prevent going astray, and they need regular feedback. However, to build commitment, your coworkers need your support: someone who listens, who not only corrects mistakes but also praises and encourages, motivates why their work is important, and shares information and insight to make them feel connected to your goals.

## 6.6 Concluding remarks

Successful teamwork is all about synergetic interactions between people who acknowledge the value that, together, they can achieve more, even if their eventual goals are not the same. Understanding people and being able to recognize their thrives and emotions are key. Being aware of the rather elementary needs of people with respect to their functioning in groups is important for all team leaders. After all, leaders probably have the same needs themselves for full enjoying their role in leading the group. We firmly believe that the "3Bs" formulated (being oneself, belonging to the group, and being valued for one's potential contributions) are important, but underestimated, factors in the success of teams. We see them as essential prerequisites for establishing trust, mutual respect, an open culture for critical and constructive exchange of ideas, and pride in achieving shared goals.

## References

Barsade, S., O'Neill, O.A.: Manage your emotional culture. Harv Bus Rev; 2016; 94; 58–66.

Blanchard, K., Zigarmi, P., Zigarmi, D.: Leadership and the One Minute Manager; William Morrow, New York, 1985.

Childs, G.: Understand Your Temperament; Rudolf Steiner Press, New York, 2009.

Deutschendorf, H.: Five ways to spot an emotionally intelligent leader, June 27, 2013. www.tnlt.com

Drucker, P.F.: Managing Oneself; Harvard Business Press, Boston, 2008.

Eysenck, H.J.; The Biological Basis of Personality; Thomas, Springfield (Illinois, USA), 1967; p 35, 39.

Goleman, D.: Leadership that Gets Results; Harv Bus Rev; 2000; 78; 78–90.

Hughes, M.D., Xu, X.J., Jenkins, P., McMorn, P., Landon, P., Enache, D.I., Carley, A.F., Attard, G.A., Hutchings, G.H., King, F., Stitt, E.H., Johnston, P., Griffith, K., Kiely, C.J.: Tunable gold catalysts for selective hydrocarbon oxidation under mild conditions. Nature; 2005; 437; 1132–1135.

Lencioni, P.: The Five Dysfunctions of a Team: A Leadership Fable; Wiley, New York, 2002.

Lee, A.: How are doctoral students supervised? Concepts of doctoral research supervision. Stud Higher Educ; 2008; 33; 267–281.
Nørskov, J.K., Bligaard, T., Logadottir, A., Bahn, S., Hansen, L.B., Bollinger, M., Bengaard, H., Hammer, B., Sljivancanin, Z., Mavrikakis, M., Xu, Y., Dahl, S., Jacobsen, C.J.H.: Universality in heterogeneous catalysis. J Catal; 2002; 209; 275–278.
Oshry, B.: Seeing Systems: Unlocking the Mysteries of Organizational Life. 2nd edition; Berrett-Koehler, San Francisco, 2007.
Rostrup-Nielsen, J.R.: 50 Years in catalysis. Lessons learned. Catal Today; 2016; 272; 2–5.
Steiner, R.: The Four Temperaments; Rudolf Steiner Press, London, 2008.
Wisker, G.: The Good Supervisor, Supervising Postgraduate and Undergraduate Research for Doctoral Theses and Dissertations; Palgrave MacMillan, Basingstoke, Hampshire, UK, 2012.
Zaccaro, S.J.: Trait-based perspectives of leadership. Am Psychol; 2007; 62; 6–16.

## Yong-Wang Li

# Guest Column: Science is the right guide for the future

*Professor Yong-Wang Li is Founding Manager of Synfuels China Technology Co., Ltd., In Beijing-Huairou.*

The climate for high-quality scientific research is not so good at the moment, I'm afraid. That is a pity, because we should all be working hard on creating solutions for a safe and sustainable future for our world. In the energy sector, for example, without fundamental scientific knowledge, we will not be able to find solutions that are truly durable. In the end, I think we will have to look at properties and behavior of materials at the yet hardly explored quantum level, to discover future energy resources that are related to the fine control of the fundamental structures and functionalities of matter due to weak interactions that we are now unfamiliar with. It easily takes 50–100 years to discover such knowledge and transfer it into technology.

For companies, such new fundamental knowledge will help us in safeguarding our competitive position, but we must be patient and we need to have courage to take a long time perspective. And we – including our shareholders – must also be prepared to invest the money we make today in definitive solutions for the far future.

It may sound contradictory, but I strongly believe that meanwhile, we must continue to use energy from fossil sources. The clean and responsible application of such fossil fuels can give us time for working on real solutions. It also generates funding for new scientific know-how and technologies and for educating young scientists for these tough challenges. We should encourage our younger generation to work in these essential and new areas.

Unfortunately, other forces exist, some push forward, but some hold us back also; many artificial regulations, outdated policies, or stupid power games have a negative influence on progress. The funding climate for academic research makes it difficult to work on long-term goals, and the too competitive climate at universities stimulates fragmentation: small projects, fast publications, and short-term successes.

In my own trust, science always gives us the right guide to the future. It is so important that we scientists take our responsibility to explain what we do, why we do it, and why we want to devote our precious time to it. The true driving force should come purely from science. In the end, it makes no sense to distinguish between pure and applied science – science is there to solve problems.

# 7 On the road to scientific (self) leadership

## 7.1 Introduction: leadership based on authenticity and agility

The path to leadership goes via self-leadership. Having experienced the journey from apprenticeship to what it takes to lead and manage your own research, to manage your own research unit, your own laboratory, or even your entire department, provides an invaluable basis of knowhow for helping others to grow in their own first career steps. But how do you become a true self-leader and a leader for others? We think that authenticity, the ability to act naturally and in line with your own personality, is a key prerequisite. The next phase is to become fully aware of what you are passionate about and how it can stimulate and help you in setting goals and realizing these. Finally, one needs to build up experience, and for this, enhancing certain skills and deepening your insight in people and processes will be needed.

The concept of authenticity is hot in the present literature on management and leadership theories (Gardner et al. 2011). Many theories on organization place strong emphasis on transparency and authenticity. People feel better when they can behave authentically and they expect to be allowed to do so in their work. Customers generally appreciate it when they are served in authentic ways, and for example the students at your institutions like to be treated, guided, and taught by professors and docents who are themselves and do not act a role or hide their true personalities behind routines that they adopted somehow or copied from role models in or outside the organization.

**3B–6T–9E**

In the world of management and leadership theories, it is not uncommon to coin concepts as the "ABC of consultancy" or "1-2-3 to Success," often to the dismay of the erudite scientist. Forgive us for dubbing our approach to (self) leadership as "3B-6T-9E concept:"

- 3B as the three requirements for authentic behavior (be oneself, be connected, and be valued)
- 6T as the six stages in the trajectory to realizing a grand plan (Triggers, Talents, Thrives, Thrills, Trails and Track)
- 9E as the nine elements for developing skills and building up experience (including skills for autonomous and team performance and devising strategies for creating high impact)

In addition, it is often said that "change" has become the constant factor in the modern society. Organizational structures in which we operate tend to change rapidly nowadays. This could be because of external factors like legislation, unexpected changes in external funding (be it positive or negative), or it could have internal causes like new appointments at the executive level (a new head of department, a new dean, a new director) or a change in the way budgets are distributed internally. This forces organizations (the department, the research group) to respond fast and have "agile" strategies, to benefit from new opportunities, or at least minimize negative effects.

https://doi.org/10.1515/9783110468892-007

Organizational units, or stakeholder groups such as teachers and students, and even individual scientists, who respond slowly are always at a disadvantage.

> *The well prepared, who constantly anticipate policy changes, and always have a plan ready, benefit most when the opportunity arises, while slow followers will find that the budget has already been spent when they come with their plans.*

Agility, being the capability to anticipate change and respond fast in the optimal way, is much easier to realize for people who behave authentically (spontaneously), certainly when they also have a clear vision of where they want to go (see also the section on dynamic strategy), realizing intentions in an ever-changing context. Rapidly changing conditions are handled best by leaders who dispose over agility, for which authenticity is an almost indispensable prerequisite. There is no time anymore for programming and deprogramming people in modified organization schemes with new guidelines prescribing how to behave.

### 7.1.1 You and your (self) leadership personality

You have, like everybody else, an inherent, natural blueprint of your four personal operational dimensions (do, think, drive, and feel). These are connected to your personality, expressed in terms of your autonomous "I" (the individual accountability), your inner "Self" (norms, values, and inner feelings of responsibility) and a third quality, which we indicate by the term "Impetus" (your inner drives, thrives, passions, and sources of inspiration). The first four, do, think, drive, and feel, largely determine your management potential, while the more personality based "I", "Self," and "Impetus" relate to your personal leadership capabilities. Table 7.1 summarizes the concept.

The "I" lives in the here and now and has a good intuition about the context and what it takes to be effective. The "I" looks at feasibility and is willing to settle for

Tab. 7.1: Personality as expressed in three personal leadership dimensions.

| "I" | "Self" | "Impetus" |
|---|---|---|
| – Individual accountability<br>– Directed at tangible achievements<br>– Willing to compromise and strike deals<br>– Egocentric side of our personality<br>– Safeguarding own interests | – Norms, values, Inner awareness of responsibilities; conscience<br>– Directed at harmony and fulfillment<br>– Social side of our personality | – Inner drives and thrives connected to external goals<br>– Passion<br>– Inner source of energy and inspiration |
| Lives in the "here and now" | Time perspective less relevant | May change slowly over time |

realistic accomplishments, for example, being satisfied when a task is completed for 80% level because it suffices for the purpose, whereas our "Self" might find it much more rewarding to do better and go for 100%. The "Self" reflects your values and responsibilities. As these are largely invariant over time, the "Self" lives in the continuity, one can also say in the "there and then." The "Self" is directed at harmony, sustainability, and wholeness. The "Impetus" is more difficult to describe but could be looked at as your inner sources of energy, enthusiasm, and passions. For many, it relates to their beliefs, or their philosophy of life, as the main source of inspiration. In essence, it relate to your authentic and free drives.

### 7.1.2 Self-leadership rests on tangible and intangible skills and capabilities

Self-leadership is both about *you* and about *what you want to achieve*. Too often, we find these two worlds apart: Managers in powerful positions often focus on performance and achieving tangible results (e.g., the drivers of Chapter 2), while trainers and coaches – "people managers" – have much more attention for the individual who has to do it based on intangible skills and personality traits.

As scientists or engineers, we have been trained to make rational and objective decisions using our cortex, the analytical and reflective part of the brain. However, many decisions in daily life are based on emotion and intuition. Personality and spur of the moment are important factors, and many people take decisions based on what they emotionally want and what feels good with them. This is a fact of life, whether we like it or not.

The art is how to integrate both the hard and tangible with the soft and intangible side, because both sides need each other and can factually empower each other. Leaders who manage to incorporate both the tangible and the intangible approach in a balanced way are much more likely to reach success with their teams, that is, realize the goals of the organization and the individual members. Think for example about the professor who, in his/her role as mentor and inspirer of great research, successfully completes a challenging research project, while the very motivated PhD students graduate with excellent theses and publications and the postdocs position themselves favorably for the next phase in their careers. Everybody wins; moreover, the university is happy, science is enriched with new knowledge and insights, and the society benefits from new talent and perhaps even new opportunities for innovation and exploitation may be found.

Self-leadership also strongly concerns your capabilities: job skills, life skills, and personal skills. Some skills come naturally to specific people, but in general, they derive from a mix of talent, experience, and having learned and enhanced them by specific trainings and courses. We are all different – some capabilities come naturally to us while others can be tough to acquire, if at all.

It is important to realize that people learn and acquire expertise in circular ways. Similar issues pop up again and again, and experience helps in dealing better with them each next time. However, in the negative mode, recurring challenges

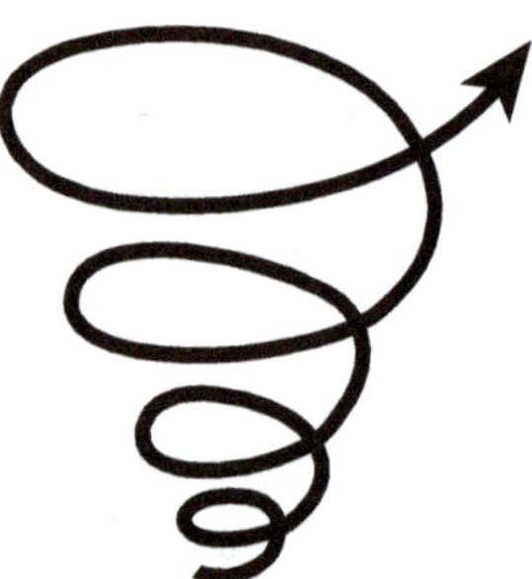

Fig. 7.1: Gaining experience is like an upward spiral – it grows and widens scope with time.

may also become a continuously growing barrier and eventually a blockade to performance, for example, if people avoid a tough or confronting solution or always defer the difficult issue to others. The positive look at circular learning patterns is to see them as an upward spiral, which widens to broader scope and reach as time progresses (Fig. 7.1).

**Enhancing Servant Leadership (ESL)**

ESL (Gardner et al. 2011) is a theory based on thorough practical research into the essentials of self-control, combining the personal dimensions that energize people with the performance and impact dimensions that lead to success. We prefer the word "enhancing" to "developing." The latter literally means "unfolding" (but what is left of an onion when you peal of all the rings?), while enhancing expresses growth. For the authors, it has factually more than one meaning. Enhancing implies enlarging your possibilities to grow, to perform, to create impact, because you combine an increasingly broader scope of experience- and insight-based sensitivity and sensibility with focused decisions and attention. It also implies consciously following up on positive emotions and standing above negative feelings, making the approach focused on getting the best out of oneself, while guiding and supporting the students and other team members working with you.

Enhancing *Servant* Leadership is focused at stimulating people to get the most out of themselves. It does not mean that the leader is the servant of his people. He/she is much more experienced and has the leading role, in letting his/her people grow while they and the organization perform. The leader is accountable and knows how to stay in control of his/her organization. However, he/she does not resort to commandeering or suppressing people; he/she leads his/her people to self-control with responsibilities and accountability within the organizational settings.

Felderhof, J.P.K.: Enhancing Servant Leadership through Inspiring, Enabling and Focusing Enhancing Self-leadership; Informance Publishers, Eindhoven, 2007.

## 7.2 Part 1. The power of authenticity and feeling recognized – the 3Bs

As we said before, authenticity plays an important role in leadership theories. What exactly is it? The adjective "authentic" means genuine, real, neither false nor copied, and representing one's true nature or beliefs, and "authenticity" is the quality of being authentic. (Kernis 2003) describes authentic behavior as "acting in accordance with one's values, preferences, and needs as opposed to acting merely to please others or to attain rewards or avoid punishment through acting 'falsely.' Authenticity is not a compulsive effort to display one's true self, but is the free and natural expression of core feelings, motives and inclinations."

People who behave authentic are as they are; they intuitively or consciously know their strengths and accept/admit their weaknesses, whether openly or not, and they are themselves under all circumstances. They do not play a role nor copy the behavior of others to make them appear better than they are – they are themselves.

What is needed to be authentic? Apparently, three aspects in life are essential for people to function authentically and all three regard mutual respect, fair treatment, trust, and recognition:
1) the feeling of being allowed to be oneself (in contrast to feeling rejected);
2) the feeling of being connected to the people who are important in one's functioning (opposite to feeling left alone or deliberately excluded); and
3) the recognition for making a (potentially) valuable contribution (with on the opposite end of the scale the feeling of an inferiority complex).

**3 B's for free authentic resources**

Be oneself – Belong to the group – Be valued

Note the word "feeling" in all three points, emphasizing that it is the subjective perception of the individual (student, postdoc, employee, or yourself) that matters and not how the boss or the environment assesses the situation.

People who are recognized by their environment extract positive energy from this feeling and encouragement to go on and do even better. Of course, recognition needs to be deserved, e.g., by showing genuine interest in the project and the people, by working hard, by being proactive, etc. Fear of rejection, on the other hand, can undoubtedly serve as a motivator for working and trying hard, but at the risk of unwanted side effects: people experiencing negative stress, trying to prove themselves, being bossy, or withdrawing within themselves and shying away from confrontation, to mention a few defensive reactions. People with a fear of being abandoned might start to counteract by avoiding conflict (e.g., refrain from critical discussion or from contributing new ideas) or even by flattering and pleasing the others. Clearly, such behavior is far from authentic. Nevertheless, we have seen many organizations, including academic groups, where the culture is fear driven, where people are afraid to make or admit mistakes, afraid to take wrong decisions or propose correct decisions that are not favorably looked upon by the management or frowned upon by the environment. Such factors work as imposed and assumed constraints (see also Chapter 2).

To avoid misunderstanding by the reader, by emphasizing the value of recognition, we do not mean that leaders should only give praise to their team or that individual members constantly give each other compliments. Such "mutual

Fig. 7.2: Comparison of career development and stages in life. At the first transition (dashed line), we see dependence disappear and maturity grow, while the dashed line on the right indicates the transition to becoming an established professional as recognized by peers. Their name, their organization, or their entire university may become a brand in the community.

admiration" societies can certainly belong to the culture of organizations, but they are hardly productive, as difficult issues and dispute are deliberately avoided to keep the atmosphere pleasant. Biting the bullet is sometimes needed, and disputes and critical assessment of ideas, provided handled in respectful manner, are essential for progress. Weak points in a research project and flaws in the execution of experiments or the interpretation of data could better be addressed immediately by the team than that they appear later and cause damage and embarrassment, e.g., during a discussion at a conference, by a critical reviewer who recommends rejection of a manuscript, or – even worse – the next funding proposal. Openly challenging opinions and critically interrogating procedures and interpretations are all feasible and in fact essential in a culture where people are respectfully acknowledged for their effort and valuable contribution.

Hence, recognition is a key enabler for feeling at ease and, thus, for authenticity, and the combination of recognition and constructive feedback (which should include criticism and points for improvement as well) is essential for high-performing teams and is particularly important for people in early stages of their career. Figure 7.2 illustrates this. A novice (e.g., student and new employee) depends largely on external recognition, until he/she has enough experience and self-confidence to realistically assess the situation and his/her own performance. The final stage on the career development scale is that of the reputed professional, who receives the recognition from peers. Each field in science and engineering has such stars, who seem to attract funding and bright students like a magnet. They have become a "brand" in their discipline.

## 7.3 Part 2. A personal plan toward a successful scientific career and (self-) leadership – the 6T approach

To be fully effective in building and shaping your laboratory, department, institute, company, or whatever it is that you want to accomplish, your personal management

Fig. 7.3: Having the four management dimensions of thinking, feeling, doing, and driving in balance is an important asset in realizing successful projects.

dimensions, thinking, feeling, driving, and doing, should come together in drawing up and realizing a plan, like in Fig. 7.3:

- To be able to imagine where you want to go and translate it into a vision, you need ***to think*** about your future.
- To be inspired and be or become enthusiastic, you need to ***feel*** the sources that give you energy and motivation. You also want to ***feel*** how trusted colleagues, students, or your environment respond to your ideas. Are they genuinely interested, skeptical, or enthusiastic? Do they believe in it?
- To make the plan work, you need the inner ***drive***, the determination and resilience to set goals and realize them.
- Finally, you simply need ***to do*** what is needed, not only the exciting try outs of novel experiments in the laboratory but also the less attractive routine activities needed to keep the organization going.

**Operational Dimensions in Teaching**

For an educator, understanding the way the student learns (*feeling*) is just as important to understand the topic (*thinking*) that somebody teaches. Compensating empathy (*feeling*) with overemphasizing the knowledge (*thinking*) and persuading the student to work hard (*driving*) may have negative effects.

This may sound obvious, but why then is it sometimes so hard to be successful in practice? If we disregard external circumstances for the moment, one needs all the four personal operational dimensions, at least to some extent. People who have brilliant ideas and great empathy to enthuse others but lack any inclination to **do** things with their hands or miss the inner **drive** to push through will probably not get very far. Or imagine the over-ambitious, perhaps somewhat egocentric driver who expects everybody to help him/her blindly, because he/she works so hard for what he/she sees as the good cause. Great thinkers may not be able to get to the execution phase, and so on. In conclusion, one needs to have the four management dimensions of think, feel, drive, and do in some type of balance.

What if we lack one or more of these qualities? The danger is not so much to dispose over too little of one or two personal dimensions; the real problem arises if a person does not recognize this or deliberately ignores shortcomings in certain aspects of management. Admitting weaknesses needs courage and self-confidence. It may well be that the environment or culture does not stimulate the person to be open and reveal vulnerabilities. A reaction that is observed often is that people hide their weaknesses and compensate deficient dimensions with a quality they are good at, in this way creating a "survivor identity."

### 7.3.1 The risk of ignoring important dimensions or declaring them irrelevant

Another complication with managers who are weak in one of the four management dimensions and – consciously or unconsciously – found a way to cope with it is that, over time, they gradually start to believe that the lacking dimension is unnecessary.

Drucker (Drucker 2008) presented this – in the meanwhile classic – example on the planner, who naturally relies on his greatly developed "think" dimension:

> *"Like so many brilliant people, he (the planner) believes that ideas move mountains. But bulldozers move mountains; ideas show where the bulldozers should go to work."*

Another example of this sort forms the scientist who compensates poor decision-making (i.e., he/she is missing the drive dimension) with overwhelmingly warm feelings to students and colleagues (the feel dimension) or developing theories of everything (the think dimension). This scientist may be popular in his/her direct environment but could have serious difficulty in securing funding or delivering output. The opposite is the hardliner, who merely relies on the do and drive dimensions and never invests in good personal relationships.

What in fact happens is that people missing out on one or more of the four management dimensions create "safe heavens" for themselves where they feel at ease. In reality, they miss out on opportunities by purposely blocking productive relations with coworkers who communicate preferentially on the dimensions that are missing.

The solution for compensating missing qualities is obviously to collaborate. An acclaimed full professor who is weak in, e.g., the feel dimension can be greatly helped by a warm and people-oriented secretary or assistant, provided both understand each other well and mutually acknowledge their roles. Willingness to listen to each other is an important prerequisite for such alliances to be successful. Another example in the academic setting is that of the visionary professor who creatively works on the fundaments of his/her field of interest and is a great motivator and mentor for his/her environment but not particularly good at finishing reports before the deadline, pushing students to finish up, or submitting proposals in time. If such a leader has an assistant professor in his/her group who focuses on these important aspects of academic life (i.e., in the do and drive dimensions), then the whole group can run very successfully. Again, a condition is that both leader and trusted right hand mutually acknowledge their roles, communicate well, and are willing to play these parts.

To realize strategies, have success, and create impact with results, one needs all four dimensions in the leadership of the unit, be it in one person (e.g., the full professor and the dean) or in the collaborative efforts of the team. In the latter case, the roles should be clear, which means that strengths and weaknesses of the players are known and mutually acknowledged and communication is open and frequent.

### 7.3.2 The 6T Trajectory to authentic leadership, high performance, and strong impact and becoming passionate about it

In our courses for personal development and scientific leadership, we find it important that participants discover what makes them passionate about their work or, if this is not what they feel, what would have to change to become passionate. The word "passion" has many meanings, but we define it here as the very strong liking of and dedication to an idea, a concept, or an activity. Being passionate is not a state of mind that one simply switches on; it grows with being exposed to the idea or being involved in the activity, and eventually, this passion becomes firmly anchored in one's inner "Self" and the "Impetus" of Tab. 7.1.

We usually start the training session by explaining some of the principles as in Chapters 1, 2, and 5, including what a vision, mission, strategy, SWOT analysis, and integral plan are. Next, we carry out a workshop where participants make a strategic plan for themselves, for their research group, or even their department. We assume that they have implicitly or explicitly a vision or at least some ideas on what they would like to achieve. We then start by showing the participants Fig. 7.4, a scheme that we have named the 6T Trajectory, meant to integrate personal plans with those of the organization. We invite the attendees to reflect a little on their current performance as a person and as a group and to be fully aware of how things are running in their environment, whether or not they are satisfied with the situation or think that there is room for improvement in whatever personal or organizational aspect of the work, including its perspective for the future. Perhaps, participants are already familiar with the SWOT analysis or an assessment based on the comprehensive scheme (Fig. 5.5 in Chapter 5), or otherwise, participants rely on their personal impressions for the moment.

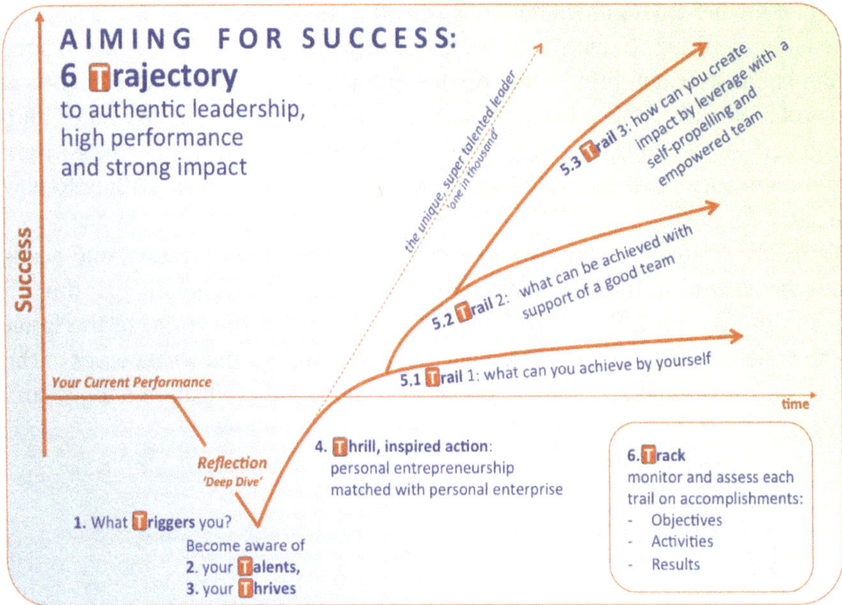


Fig. 7.4: The 6T Trajectory to authentic leadership and high performance and impact.

We then ask all participants to take a step back, and to think about what ***triggers*** them, what aspects of the work makes them enthusiastic, what gives them warm feelings, or in general, what are the sources that provide energy. Such ***triggers*** most probably reflect the person's (hidden) ***talents*** and ***thrives.*** In this respect, Freud noted that it is hard for people to go directly to the subconscious mind, but subconsciousness is often triggered by events, actualities, and situations. In other words, we sometimes surprise ourselves with hidden talents and tacit knowledge that come forward in dealing with practical situations. Once we are aware of our triggers, talents, and thrives, the next challenge is how to make optimal use of them and make them operational.

In an interview with a Dutch newspaper, Nobel Prize winner Ben Feringa was asked for his most important advice to young scientists. We translated as follows:

"Follow your dream and don't make it too easy for yourself. It is important that you discover for yourself what gives you energy. Once you know this, aim to go to the limits of your abilities. Don't be distracted and follow your intuition. This can take you very far in science."

Prof. Ben Feringa; interview by J. Jansen, Het Parool, June 20, 2017.

Now we come to the fourth and fifth T's: one's daily ***thrill*** and one's ***trails.*** To give these a stimulus and direction, one first will try to define the intention that a person, department, or organization wants to fulfill. Knowing talents and especially thrives, one can examine his/her dreams: what would one like to accomplish – in life, or, perhaps smaller, in one's working life, or, yet even smaller, in the actual job. Such an approach works for a department or an organization as well. Try to relate these dreams to talents and thrives and find out what is realistic to accomplish in the short- and long-term. Then transform the dream into an intention that is realistic and feels good. This intention should then define one's daily thrill and provide direction for the trails for achieving this Intention.

Thus, the fourth T in the scheme stands for ***thrill***, the daily feeling inspired due to realizing one's talents, thrives, and internal sources of energy. This thrill can be regarded as an entrepreneurial spirit, the inspiration for what one tries to accomplish. Identifying triggers, talents, thrives, and thrills has the purpose of bringing out the inner motivation, the reasons for feeling passionate about what one is doing or trying to accomplish. A compelling, maybe even idealistic, vision and mission form an important factor in being motivated, but if one can connect it to one's talents and thrives – in fact to one's personality – in a way that these strengths are optimally utilized, one disposes over an immensely powerful source of inspiration and energy for accomplishing goals.

If the drive derives only from external sources, say a demanding boss or environment, the person having to realize the goals may easily become frustrated or develop a burnout. In addition, without inner drive it will be hard to inspire others to cooperate.

The next phase is to sketch ***trails*** for realizing the plans. The first (trail 5.1 in Fig. 7.4) is to work out what a person can do alone, that is, autonomously, with full responsibility for execution and full accountability for end results. In the academic setting, scientists on this trail will probably be helped by students and perhaps a postdoc. Developing skills, learning by doing, and gaining expertise will all be important parts of the journey.

On the way, when the organizational unit (e.g., the research group) acquires resources, opportunities arise for trail 5.2, namely, moving in the direction of a self-propelling team. In the academic setting, one may think of the associate professor who just received tenure. The group has several PhD students, and postdocs, perhaps a technician, and especially the more senior members and the technician each contribute their own strengths as valuable assets for the entire team. This is the type of group that could well evolve in a self-propelling unit, operating like a flock of geese, where each member takes the lead on certain aspects of the work, as circumstances demand.

Some people join flocks naturally, but this is not generally the case. Before being ready to co-operate in flocks, one first should learn how to handle challenges and solve issues autonomously. The idea of a flock is not primarily to let others handle the difficulties, but to reach farther destinations together than one could do alone – all members with their own strengths and their own responsibilities. People generally like to cooperate with others who are autonomous and have something unique to contribute. Irritations about members who merely joined to benefit lead to conflicts, and negative energy, that distract from the objectives. Once people are conscious about their strengths and weaknesses, and the culture allows them to be authentic, without egos and fears blocking relationships, they can start to build on each other's strengths, talents, and natural thrives. Then co-operation in the natural way of geese flying together in flocks is possible, with the benefit to reach further with less effort.

#### 7.3.2.1 Strategic partnerships for enhanced impact

Trail 5.3 is yet another step up in potential opportunities. We have called it the strategic partnership, where teams either from the same or entirely different organizations collaborate strategically to benefit from joint expertise. The situation is probably best explained with examples. Imagine yourself in the research group of professor A from University X, with unique strengths in the synthesis of materials for a certain application, e.g., micro-electronics or molecular nano engines. At University Y, a group under a professor B has valuable expertise in computational predictions of materials properties. The two professors know each other's capabilities and decide on a strategic collaboration, where the computational expertise in Prof B's group is applied to predict materials with improved properties for the applications ventured in Prof A's group. Together, they achieve results that none of the groups separately could have obtained, and together with their PhD students and postdocs, they publish high-impact papers in top journals.

The other example, again from the scientific research front, is that of an academic group with great expertise in certain techniques or theories, which collaborates strategically with the R&D department of an industry. Together, they focus on fundamental issues underlying real-life problems from industrial practice. This knowledge enables the company to improve processes based on fundamental insight, while the academic group works on truly relevant problems and benefits from the industrial funding and from the highly relevant publications they publish in collaboration. This is synergy with high impact in action. The authors can cite several successful strategic collaborations of this type. In fact, they have been part of several such ventures themselves (van de Loosdrecht et al. 2016).

While over the years these trails are being traveled, one regularly keeps **track** of progress, evaluates whether the original goals are still relevant, or that readjustment of the strategy is in order. These examples reflected the last and fifth T (5.3) of the scheme.

### 7.3.3 Dynamic strategies

To optimize the value of collaborations, in self-propelling teams (flocks) or in strategic alliances, a shared vision is needed together with an intended and well-structured strategy. Otherwise, creating success would be a matter of pure chance, which of course may happen every now and then (like winning a lottery). Hence, a consciously intended strategy is absolutely necessary, but external factors and unforeseen

Fig. 7.5: Rapidly changing conditions urge us to almost constantly revise our strategy. Our initially defined goal becomes a point at the horizon, and we do our best to end up somewhere near. External factors cause us or give unexpected opportunities to readjust the strategy. The quality of self-leadership lends us the courage and self-confidence to take the decision of changing direction, while our sense of entrepreneurship guides us in the right direction. Self-management enables us to do everything needed to keep the project on track in the new direction, until the next point of strategy revision arises. A well-articulated vision helps to keep the project overall in the desired direction.

circumstances may render the carefully thought-out strategy useless. Therefore, strategies should be adjustable. Mintzberg already introduced the concept of emerging strategies (Mintzberg et al. 1996). Figure 7.5 illustrates the dynamic nature of an agile strategy that is adjusted when circumstances so dictate.

## 7.4 Part 3. Nine elements for enhancing your capabilities – 9Es

Third-generation governance and control incorporate insight and knowledge of all management dimensions, combined with all other sources that you have available, such as sensing, feeling, thinking, intuition, acquired experience and wisdom, creativity, and relational qualities to fully benefit from personal contacts with others, be it in formal networks or daily at work.[1]

Everybody will have their natural preferences and inclinations. The question is to what extent you have developed these qualities and which opportunities and experiences

1 Tests for scoring yourself on these dimensions with respect to personality, leadership, and entrepreneurship are available, see www.scientificleaders.com.

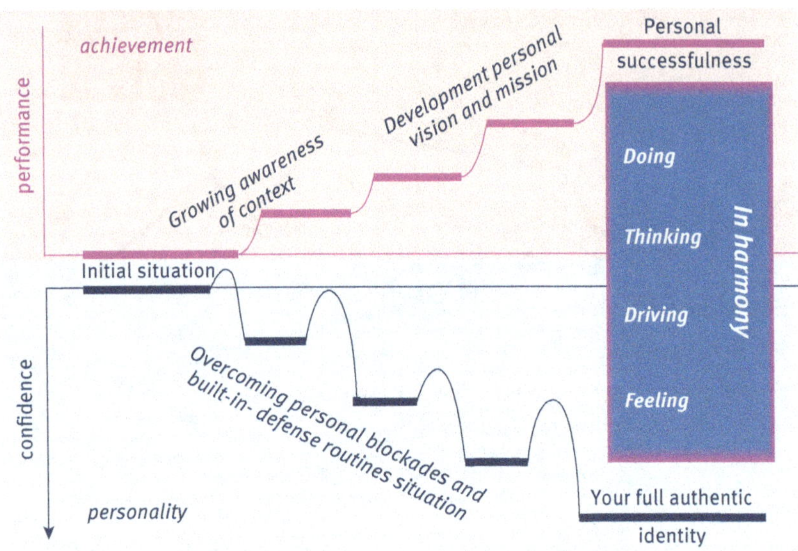


Fig. 7.6: Our success increases gradually while we gain experience and overcome personal blockades. Eventually, we manage to use our operational dimensions in a natural and balanced way that works best for us and we are confident enough to behave as our authentic selves.

you have had in your life so far to serve as nurturing ground to discover, exploit, and enhance these seven dimensions to their full potential. Learning to recognize these opportunities for enhancement is an important step toward growth. Perhaps, the situation in which you raised, studied, and acquired your experiences in life have led to a minor or perhaps more substantial unbalance, in which certain personal dimensions have been more strongly developed than others. For example, people who grew up under difficult circumstances might have developed their "I" in combination with their "do" and "drive" dimensions simply because they had to fight for their existence in a hostile environment, while others had all the chance to fully utilize their talents in music or sports, and developed strong "Impetus" and "feel" dimensions. These well-developed dimensions can undoubtedly serve as strengths in your further career. On the other hand, dimensions that received less attention could be hampering your chances to become truly successful in the future or, in other words, the chances to stay ahead of the competition.

Natural control has its roots in the seven personal dimensions we described above: all of them can be empowered and energized, stay at the present level of development, or, even worse, become frustrated, leading to blockades and feelings of dissatisfaction on your achievements. Hence, ideally, your natural dimensions should have the chance to mature toward optimal effectiveness in your daily activities (=self-management), in achieving your personal goals (self-leadership), and eventually in becoming an inspiring mentor for others (leadership), see Fig. 7.6. We propose the scheme shown in Tab. 7.2, entitled "9E – Nine Elements to Enhance Upon," as a long-term guideline for enhancing your capabilities and yourself.

Tab. 7.2: Nine elements for personal enhancement.

| Nine elements 9E | Characteristics and subjects to enhance on* | | |
|---|---|---|---|
| | **Perform autonomously** | **Perform with a team** | **Perform with teams in partnerships** |
| Full self-control | **1.1 Manage yourself** Deliver on a personal basis; invest in management skills; master self-management (based on self-management assessments). | **2.1 Lead yourself & others** Master self-leadership and start to lead others (based on (self)-leadership-styles assessment); articulate your vision and develop a strategy to carry it out. | **3.1 Lead & mentor others** Grow toward undisputed, authentic leadership at multiple levels of your organization or community. |
| Scientific entrepreneurship (= taking new initiatives) | **1.2 Playing field** Play your role in the organization; others will call upon you, develop yourself as a loyal and valued team player; explore and get to know the playing field in which you operate. | **2.2 Contact Management** Invest step by step in contacts with your peers from the wider community and start collaborations (be selective and ensure mutual benefit for all partners). | **3.2 Intentional Leadership** Convert dreams to reality in your own organizational unit (research group, department, etc.) and outside (scientific community, society in larger context, etc.; based on the personal intentional leadership assessment). |
| Scientific Innovation | **1.3 Initiate wider context** Take initiatives to bring the organization (the research group of your full professor, the department) further; take the lead in new initiatives of the department. | **2.3 Collaborate with impact** Collaborate effectively in and outside the organization; build international reputation, have comprehensive strategy for realizing your goals, learn how to build a relationship, if desirable, with different personalities than you are. | **3.3 Smart Impact Strategy** Create innovative impact by, e.g., strategic alliances, spin offs, new schools, new departments; international branches. Apply the integrated Performance and Impact Innovation approach. |
| *Examples roles* | *Assistant professor, junior research scientist* | *Associate professor, senior researcher* | *Full professor, institute director* |

* e-learning courses and assessments on several of these elements and coaching programs are available, see www.scientificleaders.com

### 7.4.1 Toward full self-control: the management aspect

The upper row in Tab. 7.2 is labeled "Self-control." Being in control over yourself is decisive for your own success and career, and therefore, it is also decisive for the success of the organization in which you are working. It is important to develop and constantly enhance your self-management roles and to learn several of the typical management skills, such as managing time and projects, handling conflicts, or becoming a master in communication. Some of these skills we have dealt with in this book, but specialized courses are usually offered by universities, and many companies organize such courses for their employees.

Self-management combines personal dimensions (do, think, feel, and drive) with performance dimensions (structure, content, people, and results). The operational dimensions stand for the sources of energy that you apply in your work, while the performance dimensions provide a focus on which you apply them in mastering your daily professional activities and achievements, see Tab. 7.3.

Tab. 7.3: The combination of operational and performance dimensions leads to 16 different management roles.

| Performance dimension | | Personal operational dimensions | | | |
|---|---|---|---|---|---|
| | | **Do** | **Think** | **Feel** | **Drive** |
| **Structure** | Carrying out well-defined operations, following protocols | *The executer* | *The organizer* | *The caretaker* | *The operational manager* |
| **Content** | Knowledge and know-how (professional, strategic, and organizational) | *The professional* | *The analyst* | *The teacher* | *The pioneer* |
| **People** | Maintaining good relationships, networking, supervising | *The networker* | *The creative dreamer* | *The social center* | *The people manager* |
| **Results** | Accomplishing goals in often complex situations, generating output | *The go-getter* | *The reflective manager* | *The challenger* | *The crisis manager* |

The combination of personal and performance dimensions yields a $4 \times 4$ matrix with 16 distinct management roles. We will not analyze these in this book but mention that in an advanced course program, assessments are available to investigate how candidates score on these roles and how they can improve on their potential. Such an assessment shows one's capability level for each performance dimension.

### 7.4.2 Toward full control: the self-leadership aspects

For the more experienced researcher, who already earned tenure at the university or is, for example, on the way to senior scientist in a research institute, self-control involves self-leadership as the fundament for balanced, successful, and endurable careers and organizations. Self-leadership is needed for being the director of your own life. In terms of the seven dimensions, you need the "I" and the "Self" in balance.

**Compromises pose challenges**

By nature, scientists score high at the content dimension. We see often, however, that the step of turning knowledge into output can be difficult. In the context of scientific research, this may well be linked to the desire to deliver complete projects, without loose ends or unanswered questions. However, reaching such a level of completeness would take an unrealistically long time, and hence, often one is forced to settle for less than ideal and be satisfied with, for example, for 90% instead of 100%. Making such compromises can be difficult and handling uncertainty is something else then dealing with analytics. It requires determination, emotional strength, and power of will, which is built upon the "I" dimension, in combination with practical experience.

Your "I" and your "Self" form a duality. One (the "I") is very satisfied if 85% is reached and the job can be considered done. The other (the "Self") strives toward perfection, toward 100%. What to choose? You need them both if you want to be successful. Your "I" takes care of your interests and position. Your "Self" takes care of relations, honesty, and continuity. So, choosing means you gain one and you lose one, but you need them both.

The duality of the "I" and the "Self" can cause big challenges in life. Nevertheless, they are like two sides of a coin; one needs both, otherwise it is not a coin.

### 7.4.3 The duality paradox between the "I" and the "Self"

People who feel insecure for whatever reason may have difficulty to behave truly authentically. They are likely to develop a mode of behavior that allows them to cope with the situation, for example, by being overly assertive or, on the contrary, rather giving in and surrendering than seeking a confrontation or even choosing to please others to avoid unpleasant situations. In such cases, the "I-Self" duality becomes a paradox that is hard to solve and becomes a source of stress, imbalance, egocentric behavior, or undesired alienation and disruption. People may not feel that they are recognized, belong to the group, and appropriately valued for their input, talents, and efforts. Such situations give rise to survival mechanisms and artificial – that is, forced – behavior.

If our survival identities and defense routines take control over the way we behave in our relations with others, we are essentially controlled by negative emotions, like fear, anger, sadness, jealousy, or revenge. Sometimes, people have developed such defense mechanisms to survive in earlier life, in childhood, during high school, or in previous jobs. People might base their personality on survival routines, which become a self-identification. If you have such members in your team, it may be quite a challenge to recognize such self-created identities and an equally difficult challenge and a long process to help them realize that behaving in accordance with one's authentic identity gives much more fulfillment.

This is where leaders can be tremendously important in helping their team members in the development to mature, authentic, and innerly free human beings: people who are accountable, responsible, and enjoying life while following and achieving relevant goals through meaningful activities.

When people are caught in the duality paradox, they escape confronting or ambiguous situations in several ways, e.g., by wanting to control (trying to be the boss), by choosing to obey (be a pleaser, or a slave), or by avoiding power games entirely, and focus exclusively on content. Under strong management, such in principle unbalanced situations can yield productive teams: the boss takes strict control over the others and everything can work relatively well in practice, although not everybody involved will be entirely satisfied with the situation.

As an example, think about university department where the chairs strictly run their own research group and take care of their part of the curriculum. The professors compete for resources by negotiating with the dean on a bilateral basis, while the monthly faculty meeting has become an arena where the professors fight for their own interests. Collaboration between groups inside the department hardly exists. Such a department can operate successfully. However, we are convinced that a department with chairs who dare to apply the type of leadership that stimulates collaboration at all levels will have much better chances to be more successful, both in education and in doing innovative research.

When people are not caught in a duality paradox, they can co-operate openly, freely, transparently, and with trust building on each other's authentic power. The duality paradox reveals itself not only within people when they are caught in a survival mode and operate in defensive modes but also in environments that have turned into a political arena, where open communication is avoided and showing vulnerability is regarded as a weakness that is up for exploitation.

#### 7.4.3.1 Does the organization encourage authentic behavior?

The culture of an organization may or may not encourage her people to behave authentic; this creates situations between four extremes, as shown in Tab. 7.4.

Situation 1 is the ideal case of an organization that encourages authentic behavior and people who exhibit it. In fact, the organization relies on the self-control and initiative of all people involved. This implies individual and collective consciousness, a trusted context, self-confidence, and the courage to be vulnerable. People who commit to the vision and strategy of the organization rely on their self-management and self-leadership capabilities and take accountability and responsibility where possible. Organizations operating in this mode represent the most powerful governance and control situation. New team members coming from an environment or culture where authentic behavior is not encouraged (situations 2 and 4 in Tab. 7.4) may initially be confused and need time and guidance to adapt.

Authentic leadership leads to balance and harmony for people within themselves, with each other, as well as with their environment. People feel recognized, happy, and valued. They will also be dedicated to the organization and its goals, which are congruent with their personal goals.

Tab. 7.4: Organizational culture and authenticity.

| Organization | Encouraging environment | Discouraging environment |
|---|---|---|
| **People:**<br>**Authentic** | 1<br>Organization can be run naturally based on authentic leadership; people can thrive on their and each other's talents, drives, and capabilities and fully realize their potential (*Flock leadership situation: authentic, free, autonomous, responsible, and natural cooperation*). | 2<br>People conscious of the situation decide if they accept it and find their own sweet way in the system and/or attempt to influence the system (in small steps) to become more open to initiative.<br>(*Hide or fight: "If you don't like the system – change it"*) |
| **Not authentic** | 3<br>Opportunity for the leadership to create an environment of trust and mutual respect where people feel at ease and eventually develop authentic behavior.<br>Leadership should be attentive and prevent that persons in the team abuse freedom, or manipulate others to their own advantage (ensure a *safe opportunity for natural growth*). | 4<br>Organizations and people in this mode are essentially in an unstable form of governance control; the environment suppresses creativity and initiative; energies are (partially) blocked; cooperation is hampered.<br>(*Lasts as long as complying yields material and/or psychological advantages for the employees.*) |

The other extreme, situation 4, represents an unhealthy and intrinsically unstable form of control, where all involved accept the situation. Control is based on power in a hierarchical and sometimes bureaucratic structure with explicit and implicit rules. People focus on their own tasks and avoid the risk of making mistakes, rather than taking new initiatives. Perhaps they realize the situation and operate carefully, or they become dominant and/or political when they want to have power and put their own interest first. In many cases, however, people might be unconscious, follow their survival mechanism, and experience the situation as normal.

Unfortunately, many organizations, also in academia, reside to at least some extent in situation 4, characterized by a far-from optimal governance and control system, with a political arena of power interfering with people's wishes to apply their natural sources of talents, wisdom, and thrives to generate useful outcome. The principle of governing is largely to control people on what they should do or accomplish. However, control, professional know-how, creativity, and productive output are not separated in a natural process. In situation 1, on the other hand, the governance principle is to give people freedom to perform and grow, each in their own way, and to lead them toward a common goal. The way to get there is cooperation, interaction, and intercreation. In nature, this is called flock leadership.

### 7.4.4 Being in control of your organization

For you as a leader, the way to control the organization in a natural way is to give your people the right context to perform and grow through own experience. The

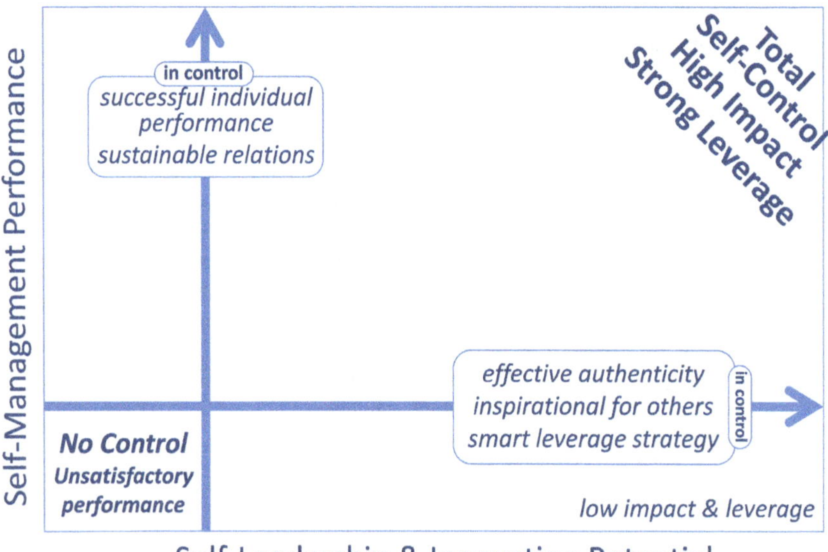


Fig. 7.7: To experience high impact and leverage, agile organizations rely on the initiative and commitment of their members, implying that both self-management and self-leadership capabilities are strongly developed.

extent to which self-management and self-leadership dimensions are developed in a person determines how far he/she is enhanced in this respect. When the degree of enhancement of a person in these dimensions is in accordance with the requirements of the function, the function and person are likely to make a good match while you and your organization have a good chance to be in control over what you want to achieve.

Figures 7.7 and 7.8 show how both the organization and its employees can be in authentic control while allowing freedom and giving responsibility. The entry level of a new employee (or PhD student) may be that of performing prescribed tasks or routine procedures appropriately (see also Fig. 6.4 in Chapter 6). Giving this person too much responsibility would not work, and both the organization and the employee would find themselves in the left upper part of the diagram. The opposite happens when somebody with well-developed self-leadership capabilities is asked to work on routine tasks. Such a person would lose interest in the job, while the organization does not benefit from the employee's potential – the bottom right of the diagram.

The ideal situation arises when employees operate at or near the diagonal of the figure, that is, in the green zone of the diagram. People and organization are both in control, and the employees have interesting and satisfying jobs, in which they are appropriately challenged.

Check – right one?

Fig. 7.8: Relation between governance and control in an organization, the level of personal development, and responsibility of the employees.

## 7.5 Flocks, teams, and flock leadership: leading on authenticity without losing control

Geese, among other birds, fly in V-shaped flocks to save energy (Lissaman & Shollenberger 1970). In flock formation, each bird flies slightly higher than the one before, to reduce wind resistance and get an uplift. The geese take turns in leading the flock and fall back when they get tired. Estimates are that they can fly up to 70% farther in large flocks than alone. They need to beat their wings less frequently and have lower heartbeat than do birds flying alone. The V-formation is also believed to be beneficial for communication and keeping the flock coordinated. In fact, professional cyclists riding in groups use similar principles to minimize the effect of head and cross winds. The V-shaped flock thus represents an ideal model of a well-functioning, self-propelling team.

**Flock leadership in the self-organized Jazz Band***

*A good example of a self-propelling team that illustrates the principle of self- and flock leadership is a jazz band, playing tunes with a lot of improvisation. Ranging from a trio to a band of any size, most of them perform without conductor, although, often, one of the musicians has the lead. They usually start with a theme, e.g., a well-known song with a fixed score for every musician. After the first verse, the musicians take turns in improvising on the theme. During their solos, they are in the lead, while the rest stick largely to their own part. Here, we see the self-propelling team of self-leaders in action, operating almost like a flock of geese in nature.*

* Thanks to Dr. Jens Rostrup-Nielsen, who often referred to jazz bands as perfect examples of well-performing teams; see page 14 for his views on leadership from the perspective of an industrial research laboratory.

By now, it will probably be evident that the authors very much support the model of flock leadership as the preferred way to lead a team, small or large, and even an organization. Flock leadership rests on a natural, authentic style and on relying optimally on the capabilities, personalities, and initiatives of the team members. Hierarchy is in place but does not dominate the culture. Power, if the word should be used at all, derives not necessarily from hierarchical position in the organization, but much more from strengths, knowledge, experience, insight, or useful relationships with external parties. In this sense, flock principles match very well with Oshry's total system power that we discussed in Chapter 6 (to avoid misunderstanding, note that "power" stands here for "power to accomplish").

### 7.5.1 How to lead on authenticity

Basically, trust is the main facilitator. Trust is important for a healthy culture with people being allowed to be themselves, which implies daring to admit weaknesses and to acknowledge vulnerability. One can be vulnerable from weakness and behave emotionally dependent on others, but one can also be "powerfully vulnerable," in the sense that weaknesses are acknowledged, while being confident and in self-control when organizing help or cooperation. In a leadership position, such people will stay in control over the situation. A culture of trust in the team is needed for people to feel assured that what happens is for the benefit of the team and its members. When people try to manipulate others in the group for their own interests or position, trust is undermined, cooperation suffers, and the team may degrade to an arena where people play games to win and gain for themselves. In such an atmosphere, people who strive for honest cooperation can easily be abused for their knowledge, creativity, and hard work, as well as openly admitting vulnerability would be looked upon as a weakness that gives others the opportunity to abuse it. Vulnerability would then imply losing position. Hence, one should take care to not let vulnerability be abused in arena cultures. However, in leading for high performance and impact, fair play and building self-confidence are essential, and warranting a safe and trusted culture must be a top priority of the formal leader.

### 7.5.2 Staying in control of a self-propelling organization

Directing on authenticity is done by giving your people space for own decisions and by giving freedom for one's own ways of working and solutions, to ensure optimal and mutual benefit from talents. We see three opportunities for exerting control without limiting your team members in the freedom to move.

- The first opportunity for control is the definition of the Playing Field. If clearly defined and regularly updated, directions and boundaries are set in which the activities and decisions take place.
- The second is to secure the trust and challenge where people can learn and strive for performance, growth, and impact. The 3B's allow the buildup of a stable organization, instead of one based on survival personalities, as discussed earlier in this chapter.
- The third opportunity for control is to match the enhancement level of your employee's personality with the requirements for the position. The self-management, self-leadership styles, and personal entrepreneurship assessments available in the online course program give not only insights and a direction in this regard, they also allow governance, control, and compliance. This is done through function assessments on the required personality level of enhancement, which can be used as a norm, which then can be matched with self-and/or 360-assessments. The self-assessments give insight in the potential and self-awareness of people and the 360 assessments a view of others what people show in practice.

The ideal of self-propelling organizations functioning in flock-leadership mode is realistic. We have seen many good examples in our own professional careers, and we are grateful to be and to have been part of such organizations. One must, however, be aware that the concept of self-propelling teams in flock mode still depends on the presence of a care-taking leader who is and remains actively involved in the activities of the team and who is constantly safeguarding that the overall direction remains that as laid out in the vision, mission, and the dynamic and therefore regularly adjusted strategy. We have also seen situations in which seamlessly functioning teams full of synergy fell back to a collection of individual projects executed by isolated people who only communicated with their functional superior or even to strictly hierarchical structures, as indicated in Fig. 7.9. Such undesirable transitions can be caused by incidents, acute conflicts in the team, lack of resources/funding due to a failed grant application, a change in leadership at a higher level in the organization, a reorganization of the organization under which the team resorts, and a multitude of other possible causes such as plain miscommunication due to people not understanding each other's minds/way of thinking. Well-running organizations are a precious commodity. Good leaders are aware of this; they will do anything to anticipate on risks and minimize their impact, to maintain their organization, their team, their department, and/or their institute in optima forma.

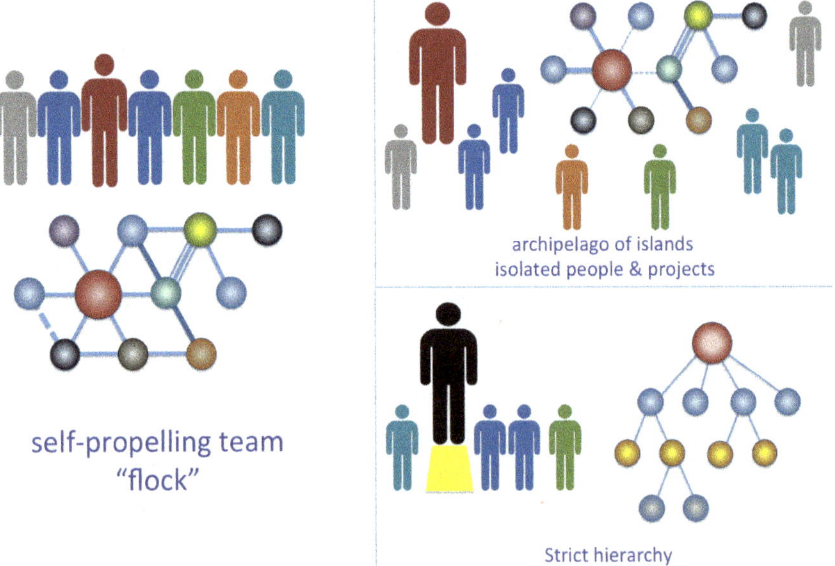


Fig. 7.9: Self-propelling teams in flock mode can, over time, decay to other modes of operation, as in the satellite system without mutual connections, other than to the leader, or a strictly hierarchical model.

## References

Drucker, P.F.: Managing Oneself, Harvard Business Press, Boston, 2008.
Gardner, W.L., Cogliser, C.C., Davis, K.M., Dickens, M.P.: Authentic leadership: a review of the literature and research agenda. Leadership Q; 2011; 22(6); 1120–1145.
Kernis, M.H.: Toward a conceptualization of optimal self-esteem. Psychol Inq; 2003; 14; 1–26.
Lissaman, P.B.S., Shollenberger, C.A.: Formation flight of birds. Science; 1970; 168; 1003–1005.
Mintzberg, H.,J. Lampel, J.W., Quinn, J.B., Ghoshal, S.: The Strategy Process, Concepts, Contexts, Cases, Prentice Hall Europe, London, 1996.
van de Loosdrecht, J., Ciobîcă, I.M., Gibson, P., Govender, N.S., Moodley, D.J., Saib, A.M., Weststrate, C.J., Niemantsverdriet, J.W.: Providing fundamental and applied insights into Fischer–Tropsch catalysis: Sasol–Eindhoven University of Technology Collaboration. ACS Catal; 2016 6; 3840–3855.

# 8 Post scriptum – implementation

## 8.1 How to implement self-leadership and the 3B-6T-9E enhancement philosophy in your organization

### 8.1.1 At the level of your team

To stimulate your team members in adopting a culture of trust and mutual respect, and to enable them to build self-confidence, they need to subscribe to the vision and mission of the team, despite that goals may vary among the team, as long these fit within the leader's strategy. An example is the win-win situation of a PhD student graduating with high marks on an excellent thesis that forms an essential building block in the overall project. A second set of conditions for people adopting flock-like behavior are the 3Bs discussed before, being one self, belonging to the group, and being valued. To achieve this, we believe that three conditions should be fulfilled:

- First, convince the members of the group that they are allowed to be as they are, without having to prove themselves. This prevents people from suffering from fear of being rejected or of not being taken seriously "because they know so little". It may take some time before newcomers feel secure in this respect, and one should be aware of strong cultural differences that may play a role with international group members.
- Second, give your people constantly the feeling that they belong to the group and have a place there. It means that they will not be excluded from important meetings or social gatherings and that their questions on things they do not know or understand yet are always taken seriously. Newcomers are part of the group, because they believe in the common goals and are willing to contribute to these. Fear of being abandoned or excluded can be strong for novices.
- Third, give your people the opportunity to grow in their own way, with enough room for one's own initiative, encourage them and show recognition for their ideas, efforts, and contributions. In such an atmosphere, it is easier to deliver critical and constructive feedback. Certainly, in the field of scientific research and development, people need freedom to move and challenges to grow through their own experience. This may imply that sometimes, the more experienced team members should hold back immediate reactions and perhaps ignore the occasional (less-critical) mistake, to give the novice the chance to discover it independently and learn. When allowable, this takes away the pressure of being looked over the shoulder continuously. Also – and again if circumstances permit – allowing people some frustration of failure has an educative effect and makes them stronger and more self-confident to find things out themselves. Learning to handle and solve complicated issues and add value to the team in their own way is a rewarding experience. At the same time, inner freedom comes with autonomy

https://doi.org/10.1515/9783110468892-008

and responsibility. The more self-confident your team members become, the more responsibility they can handle, the more autonomous they become, and the more freedom they will feel and acquire. However, offering your people safety, care, and freedom only is not enough and does not work; in return, you expect responsibility and accountability and, to a high extent, autonomous initiative. To achieve this, the best approach is probably to give trust, manageable freedom, and attainable challenges. Gradually, they can build self-confidence, take on responsibility, and gain freedom for initiative. If successful, your organization may look forward to sustainable growth and impact.

In the next phase, you should share with your team the Playing Field (with the structure, mission, vision, ambition, goals, culture, team members, and team play, etc.) in order to give them a compass for working in your environment. Ask for feedback how they perceive the implementation of your Playing Field, as this stimulates their involvement. Suggest that they make their own 6T Trajectory plan, incorporating their and your shared interests. Decide on the final 6T plan together to have it integrated with your Playing Field. This gives commitment. Then support them to perform and create impact with their 6T plans. For this, the 9 Elements program may help them.

### 8.1.2 At the level of a larger organization: the top-down approach

When you are leading a larger unit such as a department, a university, or a research institute, you will not have frequent contact with most of your people. The only way to implement principles of flock leadership is via others. You can start a top-down process and begin with sharing your ideas of governance and leadership with heads of department, group leaders, chair holders, board members, or in a hierarchical structure, the people reporting to you. In our practice with clients, we start with a presentation on flock leadership to make the participants acquainted with the advantages, trust, freedom, openness, and responsibility, as well as with the 3B, 6T, and 9E principles. Further, we use a custom-specific version of our integral governance model (like the comprehensive scheme in Chapter 5), to analyze and share the actual situation in an open atmosphere. By using the colored-notes approach (Chapter 5, Fig. 5.7), practical as well as strategic issues are brought to the table. In this way, one hopes that in addition to some successes and strengths one can be proud of, also hidden frustrations and resistances become apparent, which forms part of a healing process aimed at building trust. When done properly, people will feel energetic and pro-active after such a session. Having experienced this process in a constructive way, the participants may be willing to share their experience with the next layer and organize similar sessions in their own environment. If successful, your organization may become enthusiastic about the principles of flock leadership.

The follow-up is then similar as sketched above for a team, with the assistance of an external coach or, even better, with internal coaches, such as professionals from the human resources department, or, even better, the leaders reporting to you. Establishing cross-connections between subgroups in similar stages of development may be helpful. The cross-border element 2.3 of the 9-Elements Table (Tab. 7.2) is directed at this. You can use the triple set Self-Leadership assessments to define the required personality roles and styles for each function and match them with self- and 360 assessments. Remember, this is a top-down process. Getting all key people on the same page may prove to be a challenge, but the endeavor is definitely worthwhile if you are a firm believer in the philosophy and determined to make it work.

### 8.1.3 Middle-up-down

An alternative and slower, but in the end perhaps more durable, route to implementing the flock leadership principles in your organization is to focus on your emerging key players – the "leaders of the future". This could be in the form of an internal academy, a "Good Sports and Institute Development Program", or a "Master-Your-Future" scheme for the coming generation, such as the assistant and associate professors of a university. Participants get to know each other in a trusted environment, share experiences, and encourage each other to work along the lines of the 3B-6T-9E philosophies. The spirit, atmosphere, and their initiatives are shared with the executive level in periodical meetings, and there is a good chance that these spread organically throughout the organization.

### 8.1.4 Bottom-up

Another approach, to complement the implementation policies on a higher level, is to make the 3B-6T-9E-philosophy part of your human resources policy or human resource management development plan and target the newly arriving staff. This bottom-up approach is directed at encouraging the young staff toward self-leadership, becoming – and staying – authentic (3Bs), to make their own 6T plan and learn what is needed achieve personal and organizational goals in an effective way (9Es). This approach needs support by an experienced person, who knows what it takes to become and behave authentic, even when the context is not ready for it, to make people aware of the risks of vulnerability in an arena environment, and what it takes to handle this.

Figure 8.1 summarizes the three approaches for implementing self- and flock-leadership in larger organizations, like a University Department, or a Research Institute.

In all cases, and if desired, leaders and participants can make use of the e-learning facility at www.scientificleaders.com, where the 6T and 9E programs can be filled out, stored, and updated and where also the assessments on self-management, intentional leadership/entrepreneurship, and (self)-leadership styles are available.

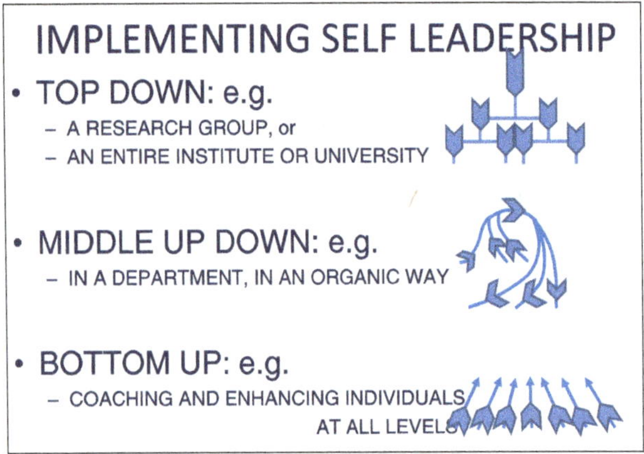


Fig. 8.1: Three ways for implementing self-leadership programs in an organization.

# Worksheets

https://doi.org/10.1515/9783110468892-009

# Score Card

## Your Personal Operational Dimensions

### to Feel

| too little | | | | balanced | | | | too much | |
|---|---|---|---|---|---|---|---|---|---|
| 1 | 2 | 3 | 4 | 5 | 6 | 7 | 8 | 9 | 10 |

Use of feelings, orientation on people, interaction, networking

### to Think

| too little | | | | balanced | | | | too much | |
|---|---|---|---|---|---|---|---|---|---|
| 1 | 2 | 3 | 4 | 5 | 6 | 7 | 8 | 9 | 10 |

Orientation on content and intrinsic meaning, often in larger context; strategy

### to Do

| too little | | | | balanced | | | | too much | |
|---|---|---|---|---|---|---|---|---|---|
| 1 | 2 | 3 | 4 | 5 | 6 | 7 | 8 | 9 | 10 |

Focus on action and execution of tasks, preferably according to established procedures

### to Drive

| too little | | | | balanced | | | | too much | |
|---|---|---|---|---|---|---|---|---|---|
| 1 | 2 | 3 | 4 | 5 | 6 | 7 | 8 | 9 | 10 |

Focus on output, achieving goals, completion of tasks, tangible results within deadline

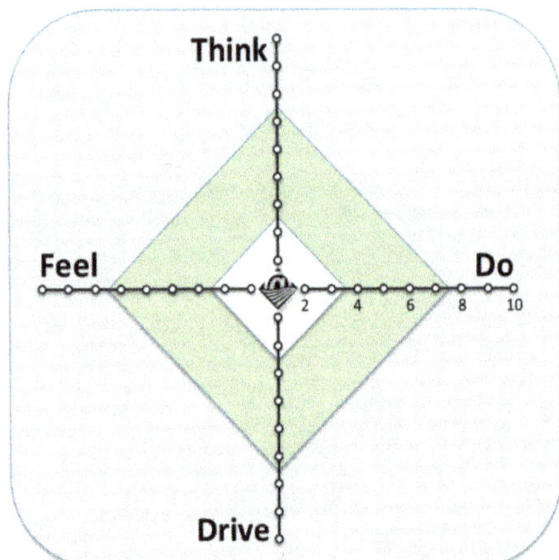

# Score Card

## Your Management Dimensions

## People

| too little | | | | balanced | | | | too much | |
|---|---|---|---|---|---|---|---|---|---|
| 1 | 2 | 3 | 4 | 5 | 6 | 7 | 8 | 9 | 10 |

The way you interact with people, supervise students, coach coworkers

## Content

| too little | | | | balanced | | | | too much | |
|---|---|---|---|---|---|---|---|---|---|
| 1 | 2 | 3 | 4 | 5 | 6 | 7 | 8 | 9 | 10 |

Your attention for content and intrinsic value of activities, background knowledge

## Structure

| too little | | | | balanced | | | | too much | |
|---|---|---|---|---|---|---|---|---|---|
| 1 | 2 | 3 | 4 | 5 | 6 | 7 | 8 | 9 | 10 |

Attention for procedures, routines, protocols, deadlines, discipline in your work

## Performance

| too little | | | | balanced | | | | too much | |
|---|---|---|---|---|---|---|---|---|---|
| 1 | 2 | 3 | 4 | 5 | 6 | 7 | 8 | 9 | 10 |

Extent to which you are satisfied with achieving goals, producing output, having impact

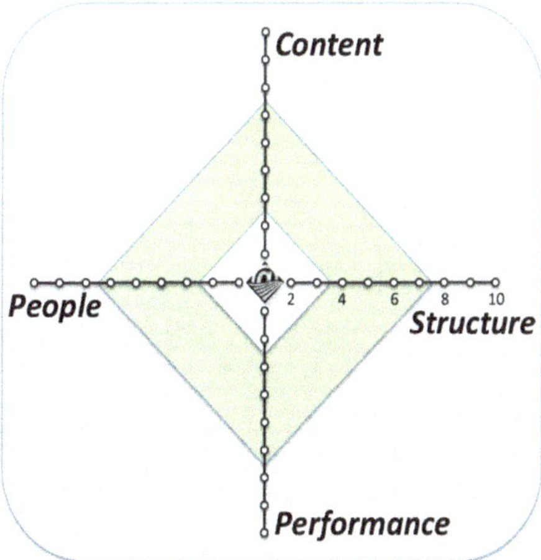

SynCat Ac@demy

# My Present Activities: 1

## An Analysis of Successes, Concerns and Failures

**Assignment:** Take 15 minutes and write down everything that comes to your mind about your activities, whether these concern daily organization or long-term planning, successes, worries about funding, issues in teaching, supervision of students, state of laboratory and infrastructure, safety, position in the Department, publications, conferences, your impact in the field, etc. Don't worry about order, or the nature of issues – small or big, just write in random order what comes up if you think about your work. Try to list at least 12 points for each category.

| What Goes Well? | What Needs Attention? | What Goes Wrong? |
|---|---|---|
| 1 | 1 | 1 |
| 2 | 2 | 2 |
| 3 | 3 | 3 |
| 4 | 4 | 4 |
| 5 | 5 | 5 |
| 6 | 6 | 6 |
| 7 | 7 | 7 |
| 8 | 8 | 8 |
| 9 | 9 | 9 |
| 10 | 10 | 10 |
| 11 | 11 | 11 |
| 12 | 12 | 12 |
| 13 | 13 | 13 |
| 14 | 14 | 14 |
| 15 | 15 | 15 |
| 16 | 16 | 16 |
| 17 | 17 | 17 |
| 18 | 18 | 18 |

# My Present Activities: 2

## An Analysis of Successes, Concerns and Failures

**Assignment:** Combine the entries in the form 'My Activities – 1' as green, orange, and red numbers (or short descriptions) under the appropriate headings in the scheme below. Is your attention for the important aspects of your work adequately spread or are there blind spots in your approach to management and leadership?

### My Organization: Why – How – What

philosophy ↕ practice

| *Vision* | *Strategy* | *'Market Strategy'* |
|---|---|---|
| *Culture* | *Management* | *PR* |
| *Projects* | *Organisation* | *Output* |

internal ↔ external

***Resources:***

| *People* | *Infra structure* |
|---|---|
| *Funding* | |

### Conclusions:

# Triggers, Talents, Thrives, Thrills

## on the trajectory to leadership

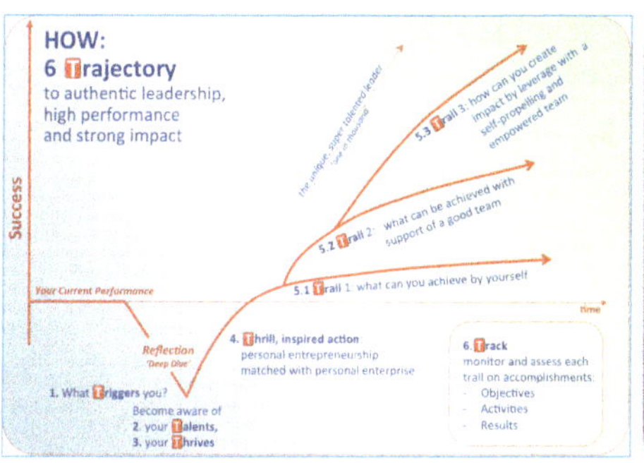


## Thrives

*What are you passionate about?*

..........................................

..........................................

..........................................

..........................................

..........................................

## Triggers

*What triggers you in your work?*

..........................................

..........................................

..........................................

..........................................

..........................................

## Trails

*What would be important steps towards success?*

*1. On your own* ..............................

..........................................

..........................................

..........................................

*2. With the help of a team* .....................

..........................................

..........................................

..........................................

..........................................

*3. With a self-propelling team?* ...............

..........................................

..........................................

..........................................

## Talents

*What are your most important talents*

..........................................

..........................................

..........................................

..........................................

..........................................

..........................................

# Team Score Card
## Management Dimensions

Score the three most important members of your team, and yourself.
Is the orientation of the team on different aspects of the work balanced?

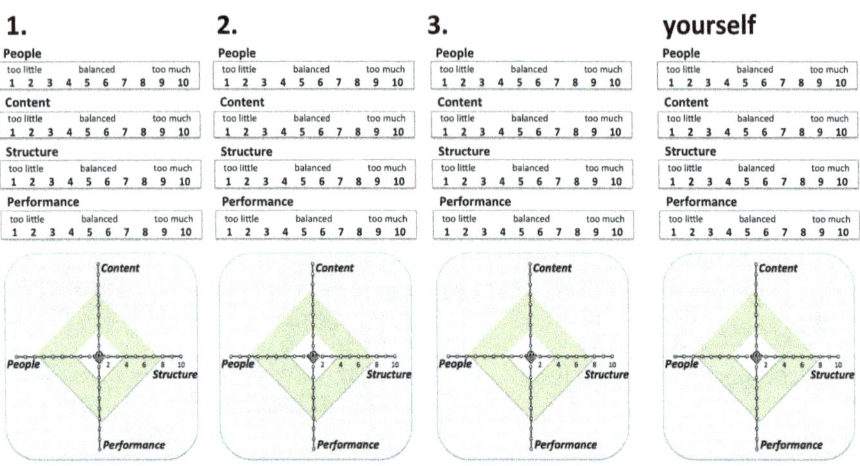


**The Team** *(all profiles combined)*:

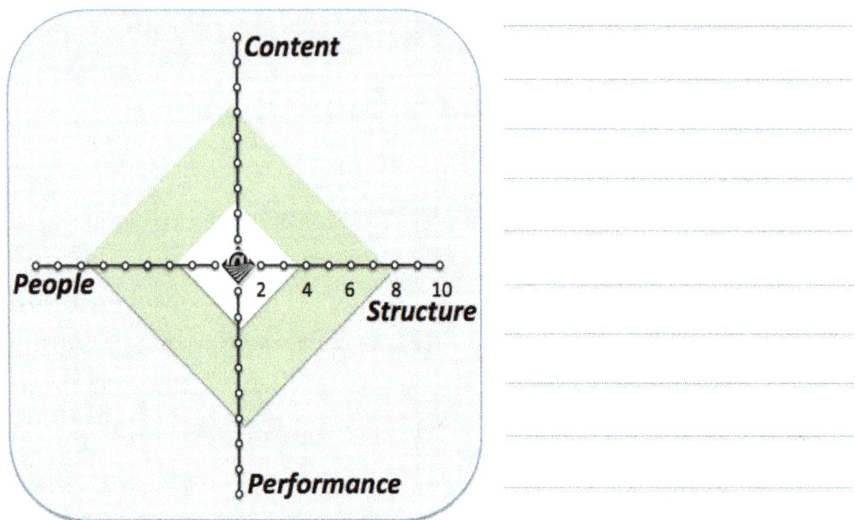

# Team Score Card
## Operational Dimensions

Score the three most important members of your team, and yourself.
Is the team balanced?

### 1.

to Feel
too little — balanced — too much
1 2 3 4 5 6 7 8 9 10

to Think
too little — balanced — too much
1 2 3 4 5 6 7 8 9 10

to Do
too little — balanced — too much
1 2 3 4 5 6 7 8 9 10

to Drive
too little — balanced — too much
1 2 3 4 5 6 7 8 9 10

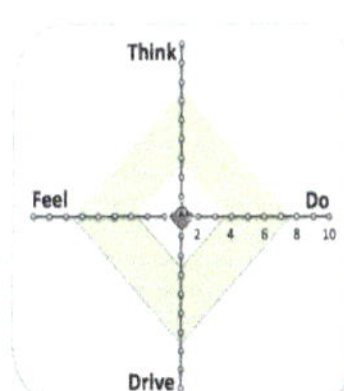


### 2.

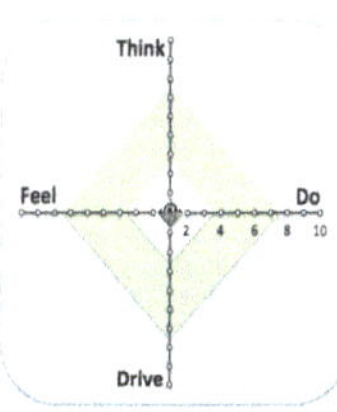


### 3.

to Feel
too little — balanced — too much
1 2 3 4 5 6 7 8 9 10

to Think
too little — balanced — too much
1 2 3 4 5 6 7 8 9 10

to Do
too little — balanced — too much
1 2 3 4 5 6 7 8 9 10

to Drive
too little — balanced — too much
1 2 3 4 5 6 7 8 9 10

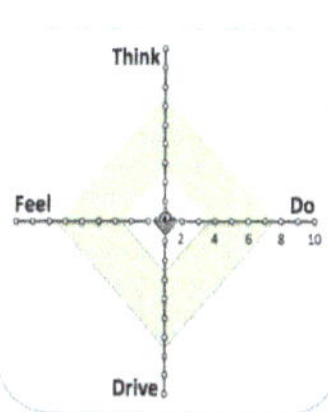


### yourself

to Feel
too little — balanced — too much
1 2 3 4 5 6 7 8 9 10

to Think
too little — balanced — too much
1 2 3 4 5 6 7 8 9 10

to Do
too little — balanced — too much
1 2 3 4 5 6 7 8 9 10

to Drive
too little — balanced — too much
1 2 3 4 5 6 7 8 9 10

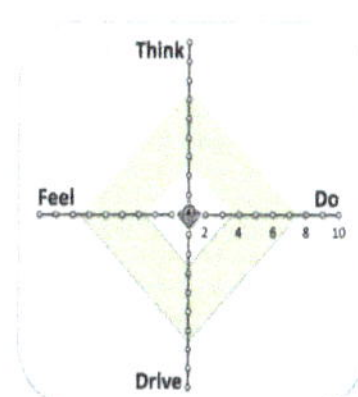


### The Team *(all profiles combined)*:

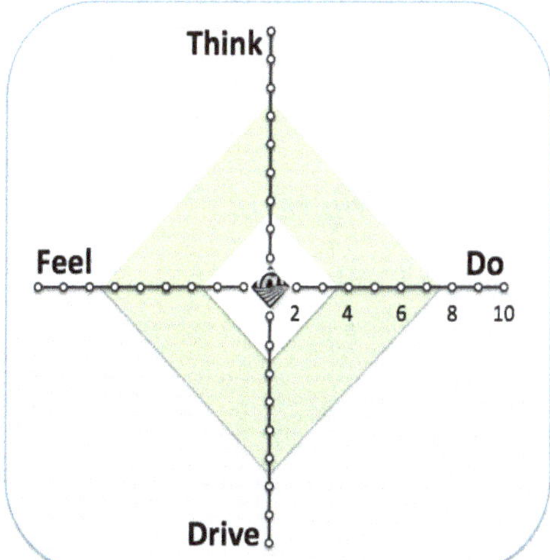

## *Vision:*

*Your view on what is needed in larger context*

..............................................
..............................................
..............................................
..............................................
..............................................
..............................................
..............................................
..............................................

## *Mission:*

*What you want to achieve in the long term*

..............................................
..............................................
..............................................
..............................................
..............................................
..............................................
..............................................
..............................................
..............................................

## *Goals:*

### *Short term*

..................................
..................................
..................................
..................................
..................................

### *Medium term*

..................................
..................................
..................................
..................................
..................................

### *Long term*

..................................
..................................
..................................
..................................
..................................

SynCat Ac@demy

# Towards a Comprehensive Plan - 2

## SWOT Analysis

| Strength | Weakness |
|---|---|
| Opportunity | Threat |

| | Opportunity | Threat |
|---|---|---|
| Strength | *Consider to go for it* | *Avoid* |
| Weakness | *Don't do it* | *Defend yourself* |

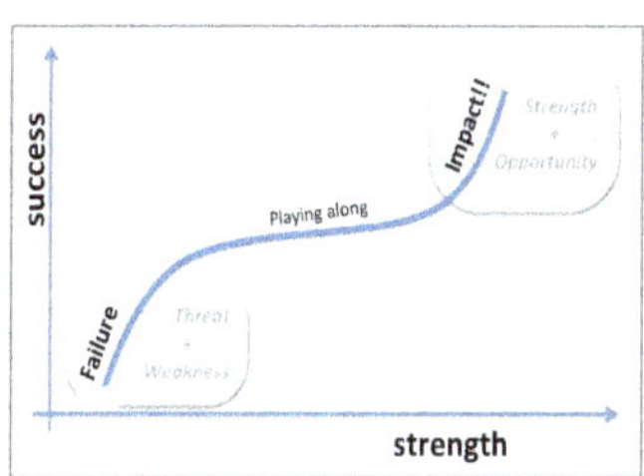

## My Organization: Why – How – What

philosophy

practice

| *Vision* | *Strategy* | *'Market Strategy'* |
|---|---|---|
| *Culture* | *Management* | *PR* |
| *Projects* | *Organisation* | *Output* |

*internal* *external*

recognition

## Resources:

*People*

*Infra structure*

*Funding*

# Index



https://doi.org/10.1515/9783110468892-010

www.ingramcontent.com/pod-product-compliance
Lightning Source LLC
LaVergne TN
LVHW081317110826
845149LV00006B/1535
* 9 7 8 3 1 1 0 4 6 8 8 8 5 *